Neues verkehrswissenschaftliches Journal

Ausgabe 26

User-based Adaptable Simulation Modelling and Design for Railway Planning and Operations

Yong Cui

2018

Habilitationsschrift

Universität Stuttgart

Herstellung und Verlag: Books on Demand GmbH, Norderstedt
Printed in Germany

ISBN 978-3-7528-5520-3

Preface

Simulation approaches are often applied to evaluate and optimise design alternatives for large-scale railway networks with a high density of train movements. However, in the past, the total potential of these simulation approaches had not been sufficiently utilised. To overcome the limits of these simulation tools and especially their incapability for expansion and customisation, a user-based, adaptable simulation tool called "PULSim" was developed.

The usability and the performance of PULSim have already been proven for existing applications; for example, the Banker-algorithm can be successfully used with PULSim to avoid deadlocks for large railway networks by searching for feasible resources. Due to its open interface, users and researchers can also develop their own algorithms that can rapidly be integrated into PULSim. With this new tool, the productivity of using simulation approaches for the evaluation and optimisation of railway planning and operations can be significantly improved.

In this book, the mechanisms and the workflow of railway simulation as well as the capability to provide users and third party applications with an open interface for dynamic interaction and flexible expansion are presented. It is not the objective of the author to build an end-all simulation tool for railway planning and operations. Instead, this developed, open interface simulation platform serves as a starting point for diverse research and development purposes that include infrastructure planning, timetabling, capacity research, and dispatching. The ability to integrate the tools needed for machine learning into PULSim especially offers great potential to improve the efficiency and productivity of railway planning and operations using the large volumes of simulation data that are now widely available.

Stuttgart, September 2018
Ullrich Martin

Table of Contents

Index of Figures

Index of Tables

Zusammenfassung

Simulationsverfahren sind eine weit verbreitete Methode im Bereich der Eisenbahnplanung und des Eisenbahnbetriebs. Heutzutage sind verschiedene kommerzielle Softwarewerkzeuge verfügbar, die nicht nur Funktionalitäten für Eisenbahnsimulationen liefern, sondern auch weitere Evaluationen und Optimierungen des Netzwerkes in Bezug auf Fahrplan, Disposition und Leistungsuntersuchung ermöglichen.

Allerdings sind all diese verschiedenen Werkzeuge in Bezug auf die Standards, die sie nutzen, sowie ihre veröffentlichten Schnittstellen mangelhaft. Für den Endnutzer sind die Grundmechanismen und Annahmen eines Simulationswerkzeuges unbekannt, was bedeutet, dass das wahre Potential dieser Softwarewerkzeuge begrenzt ist. Eines der kritischsten Probleme ist das Fehlen der Möglichkeit der Nutzer, einen ausgereiften Workflow zu definieren, der in mehreren Runden der Simulation mit variablen Parametern und Einstellungen integriert ist.

Eine nutzerbasierte, anpassbare Plattform wird in diesem Buch entwickelt und beschrieben. Die Voraussetzungen der Plattform, die Designaspekte für die Modellierung der Komponenten eines Eisenbahnsystems und Erstellung eines Workflows der Eisenbahnsimulation werden detailliert ausgearbeitet. Basierend auf dem Modell und dem Workflow wird eine integrierte Simulationsplattform mit offenen Schnittstellen entwickelt. Nutzer und Forscher erhalten die Möglichkeit, schnell ihre eigenen Algorithmen zu entwickeln, und dies mit der Unterstützung des maßgeschneiderten Simulationsprozesses, auf flexible Art und Weise. Die Produktivität der Benutzung von Simulationswerkzeugen für die weitere Evaluation und Optimierung wird bedeutend verbessert durch die offenen, nutzerangepassten Schnittstellen. Die Systemperformanz für die Eisenbahnsimulation ist von besonderer Wichtigkeit während des gesamten Designs der ereignisgesteuerten Simulation, der Berechnung von Fahrdynamiken, der Simulation von Signalsystemen, der Vermeidung von Deadlocks, sowie der Evaluation und Kalibrierung. Das Softwarewerkzeug PULSim wird implementiert, um ein einheitliches Modell, einen transparenten Workflow und offene Interfaces zu ermöglichen. Hohe Adaptions- und Performanzlevel wurden anhand mehrerer realer Beispiele mit großem Netzwerk und hoher Dichte von Zugbewegungen durch PULSim nachgewiesen.

In diesem Buch wird – zusammenfassend – die Modellierung und das Design für eine nutzerbasierte anpassbare Eisenbahnsimulation mit hoher Performanz beschrieben. Es kann als Referenz für Nutzer gesehen werden, um die Hintergründe eines Simulationswerkzeuges mit offenen Schnittstellen zu verstehen. Folglich kann ein großer Nutzen aus der Kombination von innovativer Technologie mit der Eisenbahnsimulation gezogen werden, um die Qualität und Effizienz der Eisenbahnplanung und des Eisenbahnbetriebs zu verbessern.

Abstract

Simulation approaches are widely-used methods in the field of railway planning and operations. Today, several commercial software tools are available that provide not only the functionality for railway simulation, but also enable further evaluation and optimisation of the network for scheduling, dispatching and capacity research.

However, the various tools are all lacking with respect to the standards they utilise as well as their published interfaces. For an end-user, the basic mechanism and the assumptions built into a simulation tool are unknown, which means that the true potential of these software tools is limited. One of the most critical issues is the lack of the ability of users to define a sophisticated workflow, integrated in several rounds of simulation with adjustable parameters and settings.

A user-based, customisable platform is developed and described in this book. As the preconditions of the platform, the design aspects for modelling the components of a railway system and building the workflow of railway simulation are elaborated in detail. Based on the model and the workflow, an integrated simulation platform with open interfaces is developed. Users and researchers gain the ability to rapidly develop their own algorithms, supported by the tailored simulation process, in a flexible manner. The productivity of using simulation tools for further evaluation and optimisation will be significantly improved through the open, user-adapted, interfaces.

The system performance for railway simulation is a factor of particular concern during the entire design of event-driven simulation, calculation of running dynamics, simulation of signalling systems, deadlock avoidance, as well as evaluation and calibration. A software tool PULSim is implemented in order to provide a unified model, a transparent workflow, and open interfaces. High levels of adaptability and performance have been proven by PULSim with several real examples with a large scale network and high density of train movements.

To conclude, the modelling and design for a user-based customisable railway simulation with high performance are described in this book. It can be used as a reference for users in understanding the scene under a simulation tool with open interfaces. Hence, a great benefit in combining cutting-edge technology with railway simulation to improve the quality and efficiency of railway planning and operations can be achieved.

1 Basis of Railway Simulation Modelling and Design

1.1 Simulation for Railway Planning and Operations

During the process of railway planning and operations, the most common tasks are to study the relationships between infrastructure, rolling stocks and operations, and to predict the behaviour of the studied system. Based on the learned knowledge of the system, possible solutions to improve the efficiency and the quality of railway services can be derived. Each possible solution for railway planning and operations is called a design alternative. Different design alternatives will be investigated and evaluated through experiments. However, the cost of conducting these experiments on actual railway systems is too expensive and, at times, the studied system does not even exist. Therefore, it is not practical to examine the design alternatives upon an actual system directly.

A model can be built to represent the actual system for the sake of research at a low cost. The model can be either physical or mathematical (Law 2007). Some research institutions and railway companies have built physical models in their laboratories for a certain part of railway network. A physical model represents the operational situation intuitively. It can be used for experiments of different operation scenarios or for training purposes. However, one drawback in the use of physical models is their lack of flexibility to be adapted for various railway networks in different scales. Therefore, mathematical models are more often used to describe the structure and the behaviour of the studied system quantitatively. In a mathematical model with an analytical approach, the states of the system are derived directly through a set of mathematical equations. It is practical to use analytical approaches for a simple railway network with a limited numbers of train runs or a collection of various homogeneous parts of a network in a closed-form expression. For a large scale, in non-homogeneous networks with a considerably high density of train movements, an analytical approach is not applicable due to the high computational complexity required, as well as the lack of available algorithms to link the various homogeneous parts. In this case, the simulation approach can be applied to study the structure and the interaction of a complex railway system.

The process for railway planning and operations using the simulation approach contains several steps. As the basis, the components of railway systems are modelled according to a specific design alternative. A design alternative will be simulated as an experiment, from which the output will be recorded for further evaluation. The design alternatives can be further improved in order to optimise the system performance and the quality of railway service. In Figure 1-1, an iterative process including modelling, simulation, evaluation and optimisation through the simulation approach for railway planning and operations is demonstrated.

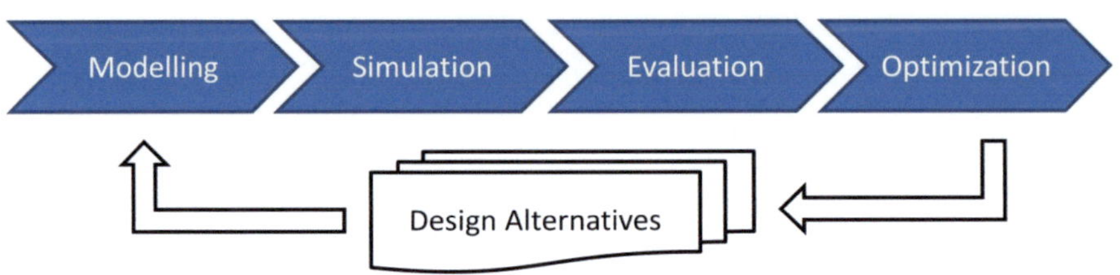

Figure 1-1: The process using the simulation approach for railway planning and operations

Today, simulation approaches are being used widely for railway planning and operations. For example, potential conflicts and resulting waiting time can be derived through timetable simulation (Luethi et al. 2005), (Radtke 2005). According to the simulated results, the path of trains and/or the sequence of trains can be adjusted in order to reduce hindrances among train runs. Furthermore, stochastic disturbances will be introduced into the simulation tool, so that the robustness and the stability of the timetable can be investigated (Radtke, Bendfeldt 2001). In capacity research, randomly generated timetables are simulated in order to evaluate system performance (Martin et al. 2011) and to identify bottlenecks (Martin et al. 2014). For re-scheduling and dispatching, the simulation approach can be applied to generate a dispatching timetable for solving conflicts during railway operations (Cui 2010). The simulation approach can also be used as a test environment for innovative technologies. For example, the potentials of using European Train Control System (ETCS) are evaluated using the simulation approach in (Kogel et al. 2010).

1.2 Design Aspects of Simulation Approach and Simulation of Train Runs

The design of simulation approaches for railway planning and operations includes two aspects: modelling the components of the railway system and building the workflow of railway simulation.

The structural perspective of railway simulation is taken into consideration in the modelling of the components of railway systems. Depending on the context, the term "model" may have many different meanings (e.g. the physical models, the mathematical models). Unless otherwise specified, the term "model" within this work refers to the modelling of the components of railway systems.

With the simulation approach for railway planning and operations, the modelled **components of railway systems** consist of:

- Infrastructure
- Rolling stocks
- Railway operations

The attributes and states of railway systems are modelled in the components respectively. The **attributes** describe the static information of railway system, e.g. the length of a track, the configuration of a train, as well as the scheduled departure time of a train run. The **states** refer to the dynamic information throughout the duration of railway operations, e.g. the aspect of a signal, the running dynamics of a train, and the delay of a train run. Normally, the components of infrastructure and rolling stocks can be constructed independently. However the model of railway operations depends on both factors.

The design of simulation workflow focuses on the behavioural perspective of railway simulation. The control flows, including starting, proceeding, and terminating a simulation process, are regulated by the **simulation workflow**. Moreover, the messages exchanged among infrastructure, rolling stocks, and railway operations are triggered inside the workflow. The model of a railway system and the simulation workflow can be designed independently. With a certain level of abstraction, the simulation workflow can be designed without knowledge of the details of the underlying model.

After the design aspects for the components and the workflow are determined, the simulation of train runs can be started, which includes the simulation of each individual train run and simulation of multiple train runs. Each train run of a given operating program (see Section 2.4.2) will be simulated until it is accomplished. Driven by the concrete mechanism of a simulation workflow, the states of the modelled components of railway systems are changed. For example, the occupancy situation of infrastructure, the aspect of signals, and the position of trains are updated during a simulation process. The calculation of running dynamics and the signalling system are the basis for updating the states of components. The interaction between infrastructure and each individual train run, as well as the interaction among multiple train runs should be taken into consideration. The changed states of the components can be recorded for further evaluation and optimisation. The relationship between design aspects of the simulation approach and simulation of train runs is shown in Figure 1-2.

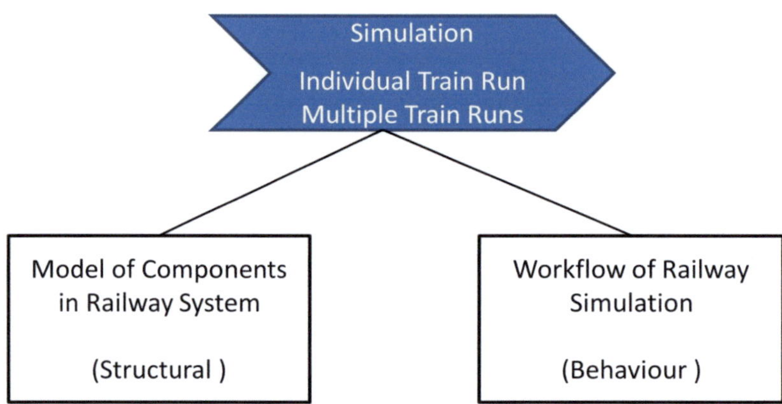

Figure 1-2: Design aspects of the simulation approach and simulation of train runs

1.3 Simulation Software for Railway Planning and Operation

Since the 1990s, serval simulation tools have been made available for railway planning and operations. Most of the tools were originally developed as experimental versions by railway research institutes and universities. Through continuous development and trial, these tools have become increasingly practical for research and commercial purposes. The user base of these tools consists of researchers, railway infrastructure companies, and railway operating companies.

In German-speaking countries and in Europe, the simulation software RailSys (by Rail Management Consultants GmbH, (RMCON 2010)), OpenTrack (by OpenTrack Railway Technology Ltd., (Hürlimann 2010)), and LUKS (by VIA Consulting & Development GmbH, (VIA Con 2011)) are very popular for railway planning and operations. These software tools provide not only the functionality for railway simulation, but also enable further evaluation and optimisation for scheduling, dispatching and capacity research. Thanks to the various applications and projects, a large amount of data of railway infrastructure, rolling stocks, and operations are now available.

However, the true potential of these software tools has not yet been sufficiently utilised. The standards and the published interfaces for the various tools are insufficient as well. For third party users and developers, it is difficult to understand the assumptions and simplifications within the used tools. Therefore, the analysis and the conclusion of a given set of simulation results can be weakened due to the lack of knowledge of the simulation tools themselves. In addition, the workload required for reading and converting data from the tools is extremely high. In many cases, the detailed information generated by the software during simulation is either unreadable or unobtainable by the user. Therefore, the usability and applicability of simulation tools are constrained. Furthermore, flexible customisation during multiple rounds of simulation for certain special purposes is also difficult to achieve. These disadvantages limit the efficiency and effectiveness of these tools for further research and development.

Especially, these tools are not yet capable of solving some well-known problems, such as deadlocks with synchronous simulation (see Section 5.2). If deadlocks are introduced, the simulation process has to be cancelled or be intervened manually. The dispatching algorithm of current simulation tools is simplified and even unknown at times. It is therefore difficult for users to verify the results and to improve dispatching processes.

For these reasons, the author has developed a simulation tool **PULSim** in recent years. In (Cui et al. 2016) and (Cui et al. 2017), the tool was known as DoSim. The name has been changed to PULSim in order to conform to the naming convention of software tools in the IEV (Institut für Eisenbahn- und Verkehrswesen der Universität Stuttgart). The aim of developing PULSim is to build a platform for railway simulation with a unified model, a transparent workflow, and open interfaces, which can be flexibly extended and customised by third parties. System performance is the key success factor for railway simulation, especially for a large network with a high amount of train runs. The main focus of PULSim is to enable the design to improve system performance and efficiency, without sacrificing accuracy. At the moment, PULSim has been tested in several real networks in Germany. In this work, the model and the algorithm used in PULSim will be presented in detail.

1.4   About the Book

Many publications about the analysis and design of railway simulation have been published in recent decades (e.g. (Radtke 2005), (Hürlimann 2010), (Waston, Medeossi 2014) and (Sipilä 2015)). Most of the research concerns the analysis of the results of railway simulation. For the design of railway simulation, the input and output, the application, and the interaction of a certain type of simulation software are focused upon herein. The detailed principles and theoretical basis for simulation design are not presented in detail.

The intention of the book is to explain the key philosophy behind railway modelling and design. Although the design decisions of software PULSim are stressed, the general principles and other possible designs are discussed and compared. The readers of the work can gain a thorough understanding of the use of simulation tools for railway planning and operations for different purposes. The results of simulation can be well verified and validated after gaining knowledge of the assumptions and constraints of a simulation model. The work can also be used as a reference to build a model for certain tasks, e.g. calculation of running time, or simulation with a certain signalling system.

The readers are not required to have an extensive knowledge of programming. The model and the workflow are presented by the Unified Modelling Language (UML). As the prerequisites for the readers, the basic understanding of railway planning and operations is, however, necessary. To maintain a unified notation and terminology, the books follows the convention and terms used in (Hansen, Pachl 2014).

In this book, the two design aspects of the simulation approach: modelling the components of railway system and building the workflow of railway simulation, are presented in Chapters 2 and 3. The algorithm for simulating an individual train run with consideration of train running dynamics and signalling systems is explained in Chapter 4. If all train runs are simulated simultaneously, conflicts may take place, and therefore the mechanism to resolve conflicts and to avoid deadlocks for multiple train runs is discussed in Chapter 5. In order to improve the plausibility and credibility of a simulation tool, in Chapter 6 the calibration methods and the practical applications are introduced. The basis for evaluation of simulation results is explained in Chapter 7. In this chapter, the open interface for simulation tools and the implemented solution in PULSim are presented as well. The design aspects in this book are based upon the recent research and development of the author and the core model used in the IEV. Besides this, certain common aspects (e.g. infrastructure modelling at a macroscopic level) that are applied also for other simulation tools are introduced into PULSim as a basis. In Table 1-1, an overview of the own design aspects, which have been implemented in the current version of PULSim, are listed and specified with the corresponding section number in this book.

Design Aspect	Section Number
Modelling	
Microscopic infrastructure modelling – infrastructure element model	2.2.4
Mesoscopic infrastructure modelling – basic structure	2.2.6, Appendix A
Definition of train formation	2.3.3
Model of microscopic timetable	2.4.3
Workflow	
Event-driven simulation	3.2.2
Individual Train Run	
Setting entry velocity and exit velocity	4.1.3
Detailed calculation of running dynamics	4.1.4, Appendix B
Simulation of train run with fixed block distance	4.2.2
Multiple Train Runs	
Identification of conflicts	5.1.2
Resolution of conflicts during railway simulation	5.1.3
Deadlock resolution with feasible resources	5.2.5, 5.3
Calibration, Evaluation and Open Interfaces	
Calibration of stochastic disturbances	6.4
Calculation of occupancy time (fixed block distance)	7.2.1
Calculation of pending time (fixed block distance)	7.2.2
Framework and integration of open interfaces with Python	7.3.3
Graphical user interface	7.4

Table 1-1: Overview of design aspects implemented in PULSim

2 Modelling the Components of Railway Systems – Structural Perspective

2.1 General Aspects in the Modelling of Railway Systems

2.1.1 Level of Detail in the Modelling of Railway Systems

In the process of modelling railway systems, the components of the system are abstracted and simplified at a certain level of detail. The level of detail for railway models depends on the specific application contexts and the purpose of simulation. The computational workloads should be taken into consideration alongside. The higher the level of detail of the model, the more information can be presented and processed. On the other hand, a highly abstracted model hides the insignificant characteristics of railway system, thus reducing computational workloads. During modelling processes, the level of detail and the abstraction level should be carefully balanced.

Three different levels of detail for railway systems: **microscopic**, **mesoscopic**, and **macroscopic**, are defined in (Radtke 2014). Although these definitions focus on infrastructure modelling, they can be further extended to rolling stocks and operations. On a macroscopic level, the components of a railway system are modelled at the highest abstraction level for a large scale of the investigated area with a very large number of train movements. A model on a microscopic level describes the most detailed information with high accuracy. It can be used to simulate railway operations with the highest level of detail. A macroscopic model is only able to provide a highly abstracted view of railway system, and a microscopic model is often constrained to be used in a limited area due to high computational efforts. Therefore, mesoscopic models are designed as a compromise. Depending on the application context, a mesoscopic level can be applied for a model that integrates both the macroscopic and microscopic levels, or for a model between macroscopic level and microscopic level with respect to its level of detail.

With a defined level of detail, the assumptions and simplifications of a model can be explicitly specified. Users can examine the applicability of the applied model with the given application context. Another advantage of defining different levels of detail is to share and apply a consistent process/algorithm for different application contexts. For example, a same path-searching algorithm can be used for microscopic, mesoscopic or macroscopic levels. With a consistent process/algorithm, a multi-scale simulation can be achieved, which integrates the microscopic, mesoscopic and macroscopic levels together inside a common process-oriented framework (Cui, Martin 2011). In Sections 2.2, 2.3, and 2.4, the different levels of detail for modelling infrastructure, rolling stocks and operations will be discussed and defined.

2.1.2 Data Model and Interfaces

The components of a railway system will be organised in a data model, which describes the contents and the relationships of the data elements. A data element represents a real entity (for example, a train or a signal) or an event between entities (for example, the triggering of the release of a block section, if a train leaves a signal releasing point).

There are three kinds of perspective of data models (Tsichritzis, Klug 1978): conceptual data model, logical data model and physical data model. A conceptual data model describes the semantics of the domain without concrete implementation details (e.g. specific database or file format). Unless otherwise specified, in this book the railway system is modelled from the perspective of a conceptual model using the UML. A logic data model represents the semantics through a particular data manipulation technology (e.g. databases, object oriented classes or other file formats). A physical data model concerns the concrete technical solutions including physical storage, performance and security. The logic data model and the physical data model of railway systems are not covered in this book. Only the key aspects and conceptual schema of railway systems are focused upon in conceptual data models.

In the field of public transport and railway systems, there are several interfaces defined for standardising and interchanging data. These interfaces can be implemented in a variety of different formats, e.g. Extensible Markup Language (XML), Comma-Separated Values (CSV), or Structured Query Language (SQL). In Europe (especially in German-speaking countries), the XML-Interface railML is widely used for railway applications. In railML, the infrastructure, rolling stocks and operations are defined as the part of infrastructure, rolling stocks and timetables respectively. At (railML 2013), an example of an infrastructure layout and the XML data in railML format is shown in Figure 2-1. In the example, there are three tracks (tracks 1, 2 and 3) connected by a turnout. The tracks are referenced with id 1, 2 and 3, and the turnout is referenced as s1. This example will be also used to demonstrate the modelling concepts in Section 2.2.

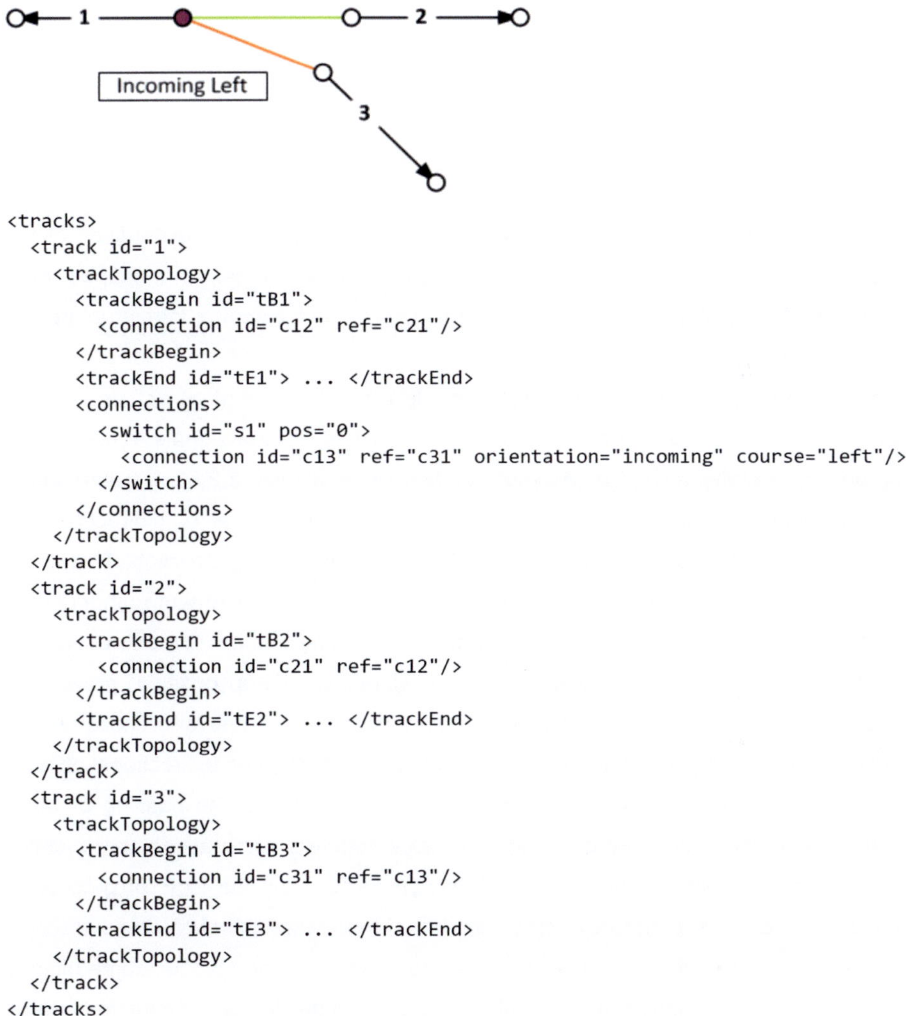


```
<tracks>
  <track id="1">
    <trackTopology>
      <trackBegin id="tB1">
        <connection id="c12" ref="c21"/>
      </trackBegin>
      <trackEnd id="tE1"> ... </trackEnd>
      <connections>
        <switch id="s1" pos="0">
          <connection id="c13" ref="c31" orientation="incoming" course="left"/>
        </switch>
      </connections>
    </trackTopology>
  </track>
  <track id="2">
    <trackTopology>
      <trackBegin id="tB2">
        <connection id="c21" ref="c12"/>
      </trackBegin>
      <trackEnd id="tE2"> ... </trackEnd>
    </trackTopology>
  </track>
  <track id="3">
    <trackTopology>
      <trackBegin id="tB3">
        <connection id="c31" ref="c13"/>
      </trackBegin>
      <trackEnd id="tE3"> ... </trackEnd>
    </trackTopology>
  </track>
</tracks>
```

Figure 2-1: XML Data in railML as an example of infrastructure layout (railML 2013)

A simulation tool may support several different interfaces. Accordingly, a parser or Object-Relational Mapping (ORM) is necessary for converting data between the persisting information and the internal data structure or object model. However, the design of the conceptual data model presented in Chapter 2 should be independent of the supported interfaces.

2.2   Infrastructure Modelling

2.2.1   Point-shaped, Line-shaped and Complex Infrastructure Objects

An **infrastructure object** is a generic term for all the objects used in infrastructure modelling. It can represent a real object (e.g. signals, tracks or turnouts) or an abstract object (e.g. paths, block sections). Depending on the characteristics of the shape, an infrastructure object may be one of the three kinds: point-shaped infrastructure objects, line-shaped infrastructure objects and complex infrastructure objects.

A **point-shaped infrastructure object** models the concerned object as a point. Some examples of point-shaped infrastructure objects include signals, route releasing points, or stations (for the macroscopic level, see Section 2.2.5). A point-shaped or a line-shaped infrastructure object can be mono-directional or bidirectional. A mono-directional object is applied for only one specific running direction. This is the case, for example, for a main signal indicating the current movement authority (see Section 3.2.2) for a train in its running direction, meaning that it is a mono-directional point-shaped infrastructure object. The main signal, in a stop position, prevents a given train from entering into conflicts with other trains on the network. Those objects having effects on two opposite directions should be modelled as bidirectional objects. For example, a pair of axle counters can be used to detect the passing of trains through a station in different directions. The axle counters and stations are modelled as bidirectional point-shaped infrastructure objects. Usually the logic of processing bidirectional objects is more complex than that of mono-directional objects. To simplify the logic of modelling and to emphasise the function of the objects, some bidirectional point-shaped infrastructure objects can be modelled as abstracted mono-directional infrastructure objects. In Figure 2-2, an example of simplification from bidirectional objects (axle counters) to mono-directional objects (signal releasing points) is demonstrated. Axle counters are used to detect the passing of a train between two points along a track. In reality, the detection of track occupancy may be implemented by a pair of axle counters (as shown in this example) or by track circuits. A pair of axle counters can detect trains passing in both directions along the track. Therefore, the axle counts A1 and A2 in Figure 2-2 can be modelled as bidirectional objects, designated with bidirectional arrows. If only the function of signal releasing is considered, and the technical details of implementation are irrelevant, a pair of axle counters can be modelled as a signal releasing point. For example, if a train is running from left to right, the number of train axles that pass through A1 and A2 will be compared. Once the counts are the same, it can be determined that the rear of the train has cleared the section ending at A2, which indicates the clearance of the block section S1→S2. Therefore, the pair of axle counters, A1 and A2, can be modelled as a signal releasing point, R, for the signal S2 as a mono-directional object (indicated with a mono-directional arrow from left to right).

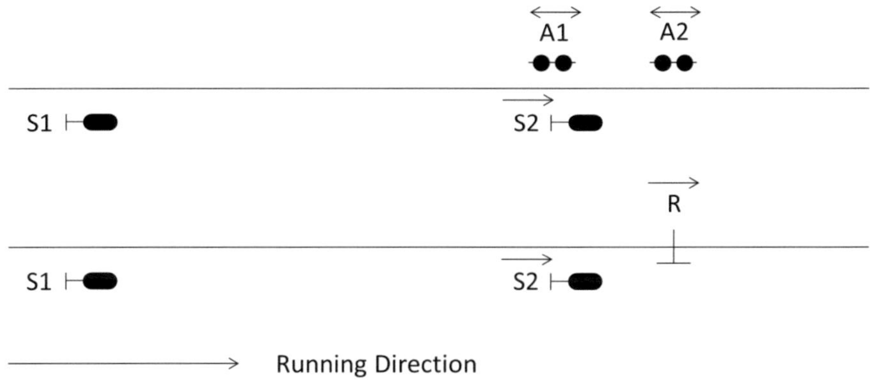

Figure 2-2: Example of simplification from bidirectional objects to mono-directional objects

Besides the direction, the coordinates and the position are important attributes of a point-shaped object. The coordinates are mainly used to demonstrate the point-shaped objects in graphical user interfaces. To calculate the distance and to search routes between two point-shaped objects, the position of a point-shaped object should be defined as the reference with respect to the whole railway network. Modelling of positions on the microscopic level will be discussed in Section 2.2.4. The coordinates and the position for a point-shaped object at the mesoscopic and macroscopic levels can be aggregated from microscopic level.

A **line-shaped infrastructure object** is connected by two or more point-shaped objects in a certain sequence. It can only consist of either two mono-directional objects or two bidirectional objects, but not a combination of the types. A line-shaped object can be mono-directional or bidirectional, depending on the point-shaped objects it consists of. The geometry of a line-shaped object is derived from the connected point-shaped objects of the line-shaped object. Examples of line-shaped objects are:

- Block section: connected by two signals, which can be main signals or shunting signals
- Path: connected by the stops of a whole train run in the macroscopic model or the exact route way in a series of blocks in the microscopic model (see Section 2.4.3)
- Path component: connected by signals/signal releasing points/route releasing points
- A straight track: connected by two endpoints
- Geometry of a curve: described by the two endpoints with its radius

A **path component** is a directed edge inside a block section used for train runs, which can be released separately as a directed occupancy element. The definition of a path component was originally elaborated in (Martin et al. 2014). A block section consists of one or several path components (if partial route releasing is applied). An overlap will be modelled as a path component for a signalling system with overlaps.

In the example shown in Figure 2-1, the (bidirectional) point-shaped objects are the endpoints with id tB1, tE1, tB2, tE2, tB3, tE3 and s1. The (mono-directional) line-shaped objects are the connections with id c12, c21, c13 and c31. In the example shown in Figure 2-2, the (mono-directional) line-shaped objects are the block sections from S1 to S2, as well as the overlaps S1 → R1 and S2 → R2.

A **complex infrastructure object** serves as the combination of the above point-shaped objects and line-shaped objects. A complex infrastructure object contains at least two line-shaped objects, with or without point-shaped objects. Examples of complex infrastructure objects are stations (at the microscopic level) or infrastructure elements (tracks, turnouts, crossings, single slips or double slips, see Section 2.2.4). A complex infrastructure object may not have a specific direction, since it consists of several infrastructure objects with diverse directions. A complex infrastructure object can also consist of several constituent complex infrastructure objects, although it can be decomposed into point-shaped and line-shaped infrastructure objects eventually. For example, a station modelled at the microscopic level as a complex infrastructure object contains several infrastructure elements (e.g. tracks, turnouts and crossings).

It makes sense to present the station as a combination with several complex infrastructure objects (turnouts and crossings), as well as point-shaped and line-shaped objects (e.g. tracks, main signals, stops and block sections), in order to show the hierarchy of the entire object. An overview of point-shaped, line-shaped, and complex infrastructure objects is shown in Figure 2-3.

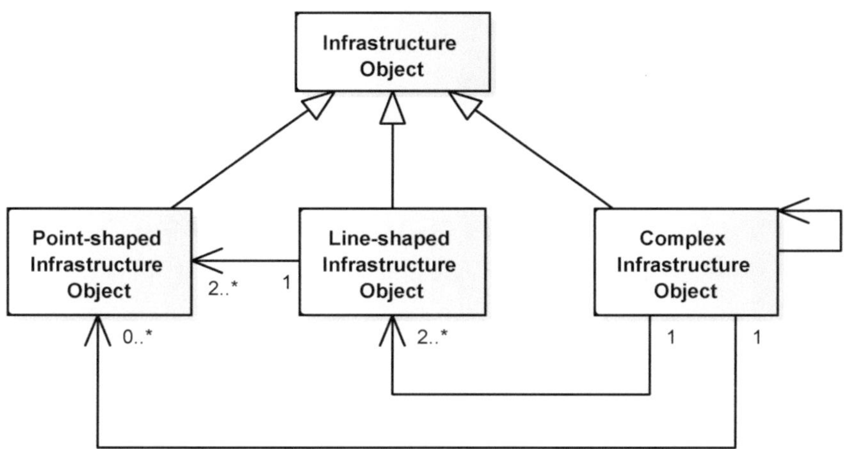

Figure 2-3: Overview of point-shaped, line-shaped and complex infrastructure objects

2.2.2   Node-Link Model

In railway infrastructure modelling, one very practical model is the link and node model (Radtke 2014), which is also often used in the graph theory.

A node represents a point-shaped infrastructure object. For example, depending on the scale of the model, a node can represent a station, turnout, crossing, or signal. A node can also represent some special positions in an infrastructure network, such as points for changing maximum allowed speed. A link represents a line-shaped infrastructure object, which establishes the connectivity between two nodes. In railway infrastructure modelling, links can be mono-directional or bidirectional. In a node-link model, nodes and links form the whole railway infrastructure network. An example of a node-link model is shown in Figure 2-1. The turnouts and signals are represented by nodes, which are connected by links.

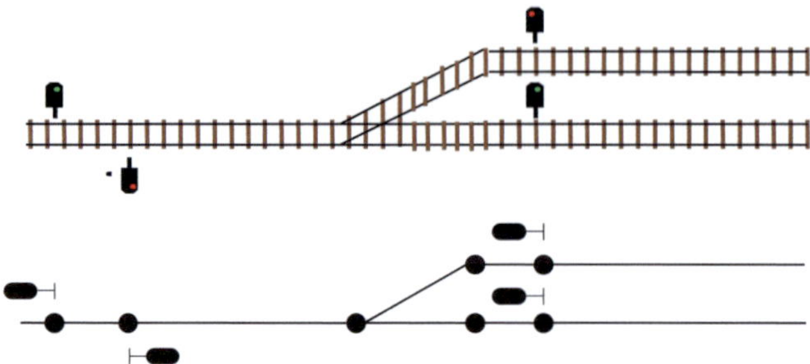

Figure 2-4: An example of a node-link model

There are several advantages of the node-link model. It is easy to learn and to implement. Since it is widely used in the graph theory, several well-known software and interfaces, e.g. RailSys and railML, are compatible with the node-link model.

However, there are some disadvantages of the node-link model at the microscopic level:

- It is impractical to model the possibilities of train runs at complex infrastructure objects.

- The effected direction of an infrastructure object cannot be modelled intuitively.

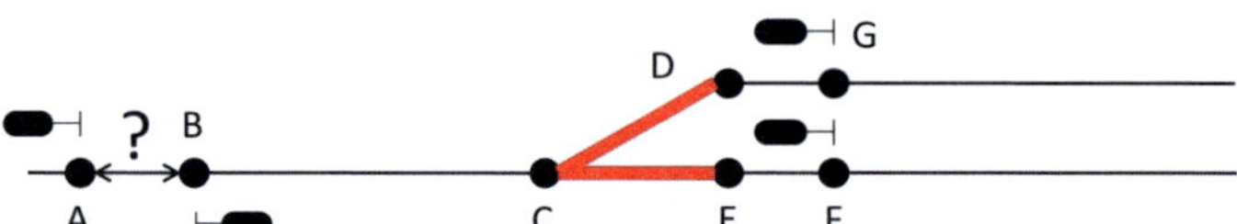

Figure 2-5: Disadvantages of the node-link model

In the example shown in Figure 2-5, a train is not allowed to run from point D via the turnout at point C to point E (marked in red colour) directly. However, it is not explicitly forbidden in a node-link mode without specifying additional rules. In addition, the effected direction for the main signal at point A is from B to A, and the effected direction for the main signal at point B is from A to B. For the link from A to B, it is not intuitive to indicate the effected direction of the link, since it has to take into account both signals at the points A and B.

Certain special designs have been developed to overcome the disadvantages of node-link models at the microscopic level. The colon-graph model (Radtke 2014) enables only the correct train runs through the duplication of the nodes. The example shown in Figure 2-5 can be converted to the colon-graph model (Figure 2-6), in which a train can only take the feasible path following the rule of "2 nodes – 1 link – 2 nodes". The path from D to E via C (2 nodes – 1 link – 1 node – 2 link) is therefore not allowed. Although the colon-graph model can be used to ensure the correct train runs, the applied rule of "2 nodes – 1 link – 2 nodes" is not intuitive for railway infrastructure.

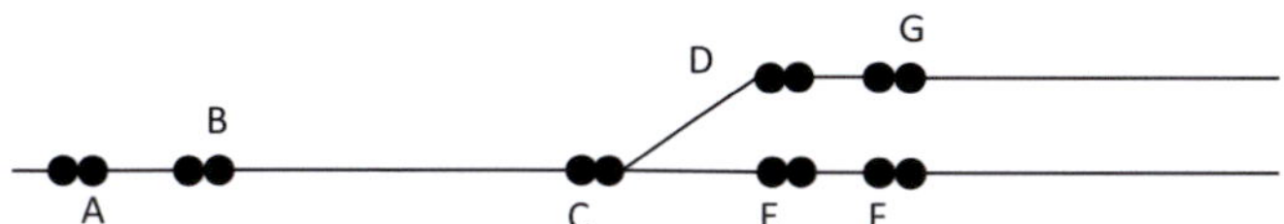

Figure 2-6: An example of the colon-graph model

2.2.3   Model of Infrastructure Elements – IEV Core Model

An approach based on an **infrastructure element model** was firstly developed in (Melle et al. 2006), which was further developed in the core model used in IEV, Universität Stuttgart (Wörner, Cui 2008) and the simulation tool PULSim. The infrastructure elements, including tracks, turnouts and crossings etc., are used as the fundamental building blocks in the model. The model is adapted closely to the nature of railway infrastructure networks. Inside an infrastructure element, each endpoint is modelled as a port, and two ports are connected via several directed edges, which indicate the possibilities of train runs.

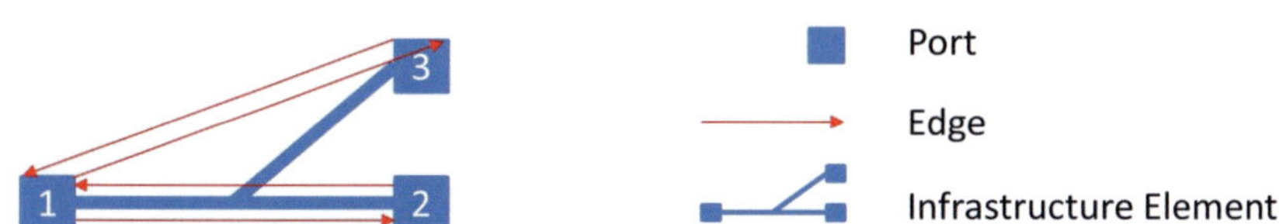

Figure 2-7: An example of modelling infrastructure elements

An example of modelling an infrastructure element (a turnout) is shown in Figure 2-7: In this example, three ports (marked with port numbers 1, 2 and 3) are modelled at the endpoints of the turnout. Four directed edges (1→2, 2→1, 1→3 and 3→1) indicate all the possible directions of train runs respectively. A directed edge will be named after the endpoints of the infrastructure element. For example, in Figure 2-7 the directed edge 1→2 in the turnout allows a train to run from port 1 of port 2. There is no directed edge connecting port 2 and port 3. It indicates that a train is not able to traverse the turnout from port 2 to port 3.

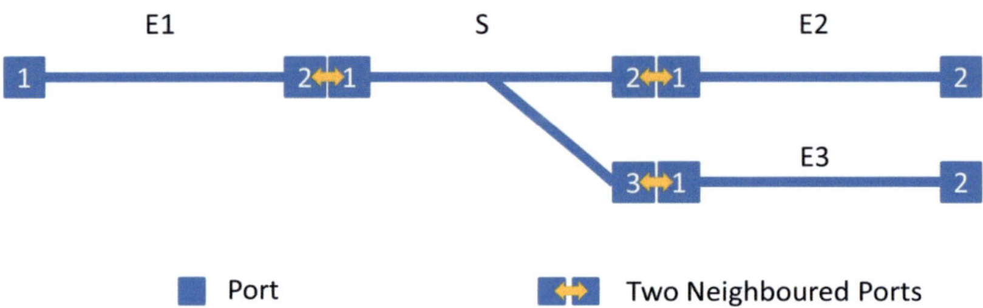

Figure 2-8: Connectivity of infrastructure elements

The connectivity of the infrastructure element model is built by specifying the neighbourhoods of the ports. The example shown in Figure 2-1 is used to build an infrastructure element model, as illustrated in Figure 2-8. Here port 2 of the track E1 and port 1 of the turnout S are neighbours. Port 2 of the turnout S and port 1 of the track E2, as well as port 3 of the turnout S and port 1 of the track E3 are also neighbours. Hence the connectivity among E1, E2, E3 and S is modelled.

Furthermore, an infrastructure element can be defined as either **non-junction-type** or **junction-type**. A non-junction-type element has two ports, and a junction-type element has more than two ports. A turnout, for example, possesses three ports and is therefore categorised as a junction-type element. On a junction-type element, two train runs may merge and/or cross from different directions.

Within the infrastructure element model, the path of a train run is not allowed to pass the same junction-type infrastructure element twice in succession. Through this simple rule, the possibility of train runs can be modelled correctly and intuitively. In the example shown in Figure 2-8, if a train run passes the edge 2→1 at the turnout S, it is not allowed to pass the edge 1→3 immediately afterwards. Therefore, the infeasible path from port 2 to port 3 via port 1 at the turnout S can be avoided.

It is also possible to apply the infrastructure element model on the macroscopic and mesoscopic levels. However, in a mesoscopic model and at macroscopic levels, a node or a link represents the aggregated information derived from the microscopic level. For this reason, it is very common that a node or link has effects for all the possible directions. Therefore, the node-link model is suitable for modelling infrastructure network at the macroscopic and mesoscopic levels (see Sections 2.2.5 and 2.2.6). The infrastructure element model has the advantage of providing detailed and reality-oriented modelling at a microscopic level, which will be presented in Section 2.2.4.

2.2.4 Infrastructure Modelling at a Microscopic Level

At the microscopic level, the model contains the most detailed information of the investigated infrastructure network. Based on the infrastructure element model, five different types of infrastructure elements (tracks, turnouts, crossings, single slips and double slips) are shown in Figure 2-9:

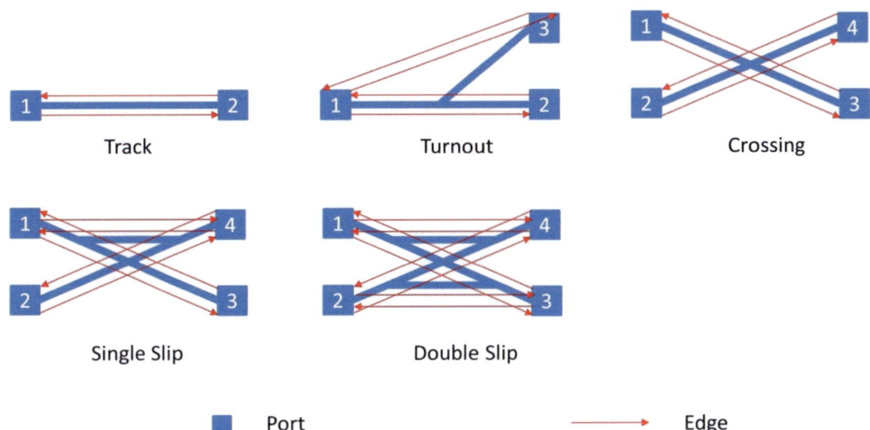

Figure 2-9: Types of infrastructure elements

A track is a non-junction-type element, while turnouts, crossings, single slips and double slips are junction-type elements. The type of infrastructure element is an important attribute used in modelling at macroscopic and mesoscopic levels and dictates the workflow of train movements.

As discussed in Section 2.2.2, an infrastructure element is a complex infrastructure object. Each infrastructure element has ports and directed edges and the possibilities of train runs are defined within it. For example, when compared to crossings, a single slip has the same amount of ports and two additional directed edges (1→4 and

4→1) .The connectivity between the two infrastructure elements is achieved through the neighbouring ports.

The point-shaped infrastructure objects at a microscopic level can be mono-directional (for example: main signals, signal releasing points, stops) or bidirectional (for example: station boundaries, measurement points of running time). The position of a point-shaped infrastructure object is defined by:

- The directed edge that the object belongs to
- The distance from the starting port of the directed edge to the object

For a mono-directional infrastructure object, the directed edge of the position indicates the affected running direction of the infrastructure object. In Figure 2-10, the position of the main signal A is referenced by the directed edge 1→2, and the position of the main signal B is referenced by the directed edge 2→1. Hence signal A will guide the train movements from port 1 to port 2. Similarly, the train movements from port 2 to port 1 are guided by the main signal B. The distances of the positions for the main signals A and B are s1 and s2 respectively.

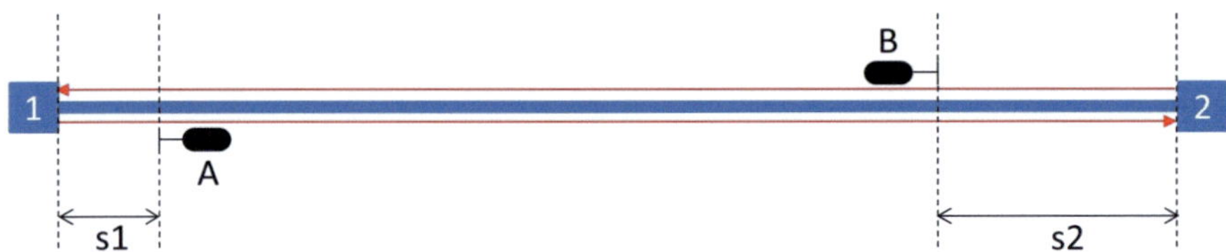

Figure 2-10: Position of the infrastructure object

Based on the defined point-shaped infrastructure objects, the line-shaped mono-directional infrastructure objects can be further created, and are referenced by starting and ending point-shaped infrastructure objects. For example, block sections and overlaps can be established after main signals and signal releasing points are created.

For bidirectional point-shaped infrastructure objects, the position of the objects can be referenced by both directions, depending on the concerned direction. A line-shaped object can also be bidirectional. In this case, the referenced starting and ending point-shaped objects can be swapped according to the concerned direction.

For each directed edge, the geometrical characteristics are modelled as geometry objects. The position of a geometry object is referenced by:

- The starting distance from the starting port of the directed edge to the starting position of the geometry object
- The length of the geometry object

Geometry objects play an important role in the calculation of running time. Different types of geometry objects are defined as follows:

- Curvature: describes the geometry characteristics for a curve, which can be a straight track, a curve with fixed radius or a spiral. For a straight track, the radius of the curve is infinite. The radius of a spiral changes linearly between the connected curve and straight track.
- Gradient [‰]: the gradient can be constant or be changing linearly.
- Superelevation: can be constant or can vary linearly.

A list of geometry objects will be created to store geometrical information for a given directed edge. In the list, the geometry objects are sorted by their respective starting distances. It is mandatory for a directed edge to maintain a list of curvature, from which the length of the edge can be derived.

The infrastructure network at a microscopic level is modelled through infrastructure elements, point-shaped and line-shaped infrastructure objects, as well as the geometry objects. The overview of the microscopic infrastructure model is shown in Figure 2-11:

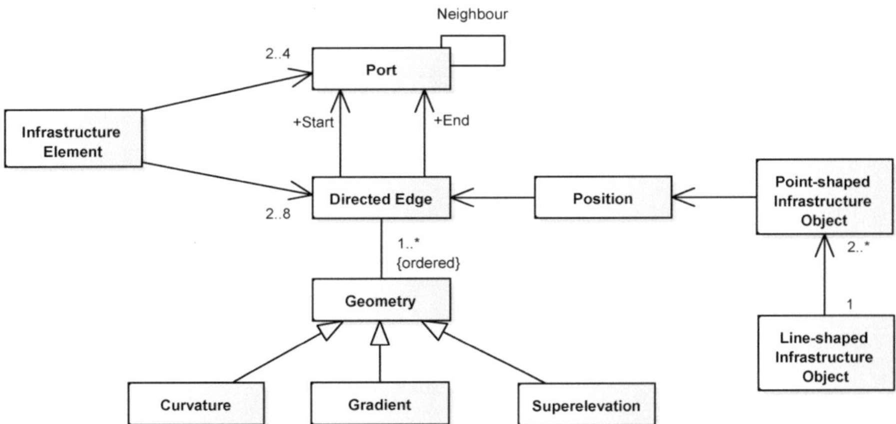

Figure 2-11: Microscopic infrastructure model

2.2.5 Infrastructure Modelling at a Macroscopic Level

At a macroscopic level, the node-link model is used to model an entire network, a subnet or a large area with the highest level of abstraction. In a macroscopic model, stations (see Section 2.4.3) and junctions (UIC 2013) are defined as nodes. Two nodes are connected via a link.

A node at the macroscopic level represents a complex infrastructure object (a station or a junction), which contains at least one turnout or crossing. Depending on different application contexts, the nodes can be further categorised by their attributes. For example, it is important to distinguish whether a train can overtake or pass another train within a node in a dispatching system. Therefore, the nodes will be categorised as loop nodes or junction nodes, with the actions of overtaking and passing only being possible within a loop node.

The term "**open track section**" can be used to specify a link at the macroscopic level. An open track section provides connections between two nodes. In case one side of the open track section terminates in a dead-end of the network, it will then connect only to one node. Depending on the signalling system and operation conditions, an open track section can be mono-directional or bidirectional. Inside a node, the possible directions of paths can be determined based on the directions of the connected open track sections. An example for mapping from the microscopic level to the macroscopic level is shown in Figure 2-12.

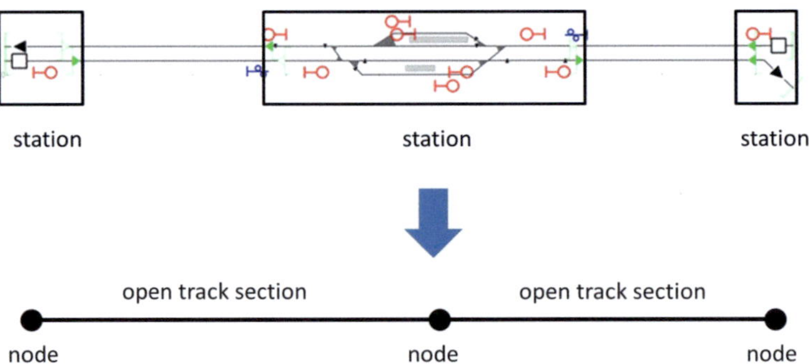

Figure 2-12: Mapping from the microscopic level to the macroscopic level

A macroscopic model can provide an abstracted overview of the investigated network with limited levels of detail, which makes it suitable for simulating a large amount of train movements in a large network. However, a validation of the abstraction should be carried out. For example, in (Cui 2010), the possible train movements on bidirectional open track sections are checked in order to avoid invalid train sequences, which may lead to deadlocks in a macroscopic simulation.

The attributes of a macroscopic model can be derived from the aggregated value based on the microscopic model. Some attributes, for example, the length of an open track section (link), can be summed from microscopic model without ambiguity. Other attributes, e.g. the speed or the curvature of a link, should be derived depending on a certain aggregation strategy. It is not always practical to simply apply the aggregated attributes at the macroscopic level. In (Radtke 2005), the variations for running time calculation with different options (lowest microscopic speed limit and highest microscopic speed limit) are compared. The difference between +6% and -20% indicates that an accurate running time calculation is difficult to achieve at a macroscopic level.

2.2.6   Infrastructure Modelling at a Mesoscopic Level

Mesoscopic models are sometimes referred to as the synergy of both macroscopic level and microscopic level models. In this book, the mesoscopic model is only designated for that level of detail between the microscopic and macroscopic levels. A node (a station or a junction) at the macroscopic level can be further modelled as a set of junction-type basic structures and station tracks (only for stations).

A **basic structure** is defined as a basic occupancy element in which all parts should be equally occupied, regardless of the design of the operating program (see Section 2.4.2). Therefore, it is prohibited for two or more trains to run simultaneously upon a basic structure. The definition of a basic structure was originally used for capacity research, in which a basic structure can be used as the basic unit to calculate occupancy time and occupancy rate (Martin et al. 2014). Similar to infrastructure elements, basic structures can be further categorised as junction-type and non-junction-type. A junction-type basic structure contains at least one junction-type infrastructure element, while a non-junction-type basic structure only contains a track or part of a track. An example of the divided basic structure is shown in Figure 2-13. The non-junction-type basic structures are designated in grey, and the junction-type basic structures are marked with various other colours.

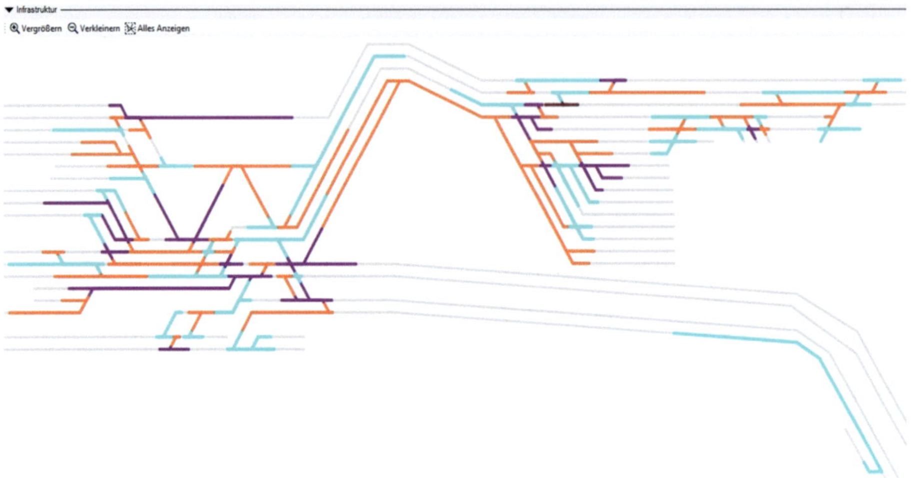

Figure 2-13: An example of divided basic structures at a mesoscopic level

Station tracks connect one or more junction-type basic structures inside a station. In the case of junction-type basic structures and station tracks, a station or a junction can be represented in the form of a node-link model at a mesoscopic level, wherein the nodes of a mesoscopic model represent junction-type basic structures, and the links represent station tracks (see Figure 2-14).

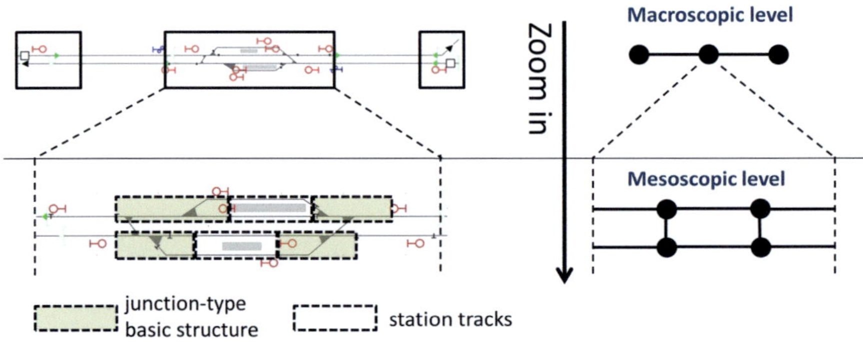

Figure 2-14: An example of junction-type basis structures and station tracks

An open track section in a macroscopic model can be modelled as one or more station tracks at the mesoscopic level. The number of modelled station tracks of an open track section depends on the number of junction-type basic structures within the connected stations/junctions. An example of the model is shown in Figure 2-15. The open track section between stations A and B contains two station tracks, which connect the basic structures A1 and B1, as well as A2 and B2 respectively. The open track section between stations A and C contains only one station tracks, which connects the basic structures A2 and C1.

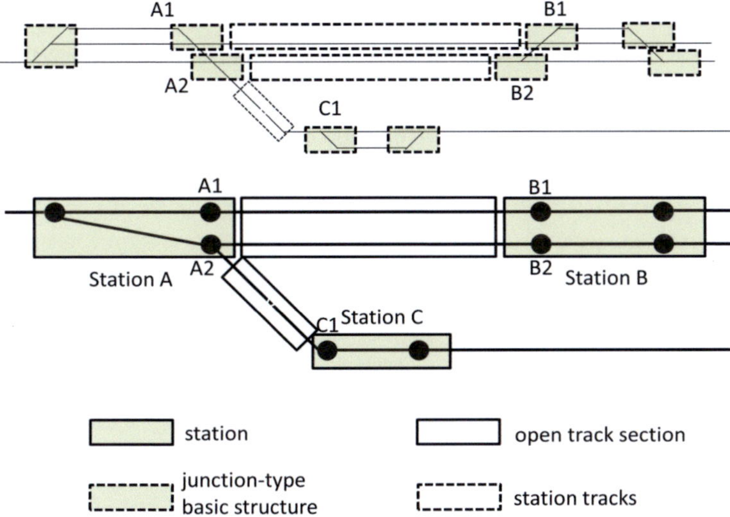

Figure 2-15: An example of open track sections and station tracks

As previously elaborated, it is not allowed that two or more than two trains occupy a basic structure simultaneously. Therefore, algorithms to automatically establish junction-type basic structures and station tracks at the mesoscopic level are necessary. A consistent partition method, which can be applied to decompose both junction-type and non-junction-type basic structures was developed by the author (see Appendix A). This method takes into account not only the network layout, but also the railway signalling system. The developed method has been applied in the project RePlan (Martin et al. 2014) for the creation of basic structures.

2.2.7 Sequence of Establishing an Infrastructure Network with the Developed Model

The process of establishing an infrastructure network starts here at the macroscopic level. The initial step will be to create stations and at this moment, only the general information including the name and the identity of a station are introduced.

Stations will be further modelled at the microscopic level. The infrastructure elements belonging to a station are then created, the connectivity of which should be assigned immediately after their creation. Point-shaped infrastructure objects (e.g.: main signals and stops) are inserted to infrastructure elements, respectively, and are referenced by a directed edge. Finally, line-shaped infrastructure objects, which connect at least two point-shaped infrastructure objects, can be built. The complete sequence of establishing an infrastructure network is shown in Figure 2-16.

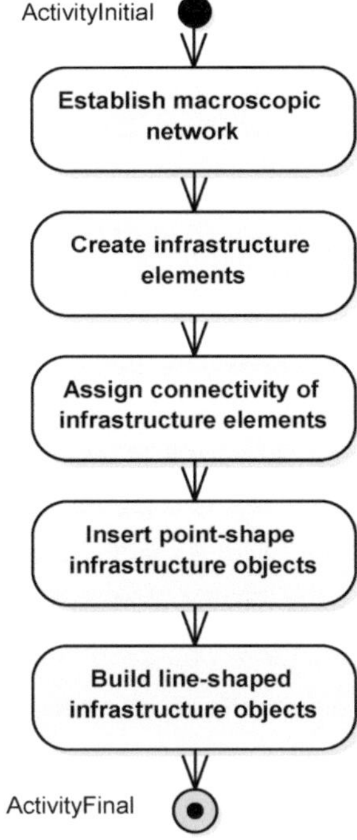

Figure 2-16: Establishing an infrastructure network

Once the information at the macroscopic and microscopic levels is available, mesoscopic modelling can be carried out. According to the application contexts and requirements, mesoscopic models are optional for some simulation tools.

The dependency among different levels can be concluded also from the sequence of establishing an infrastructure network. The changes at the macroscopic level will influence both the microscopic and mesoscopic levels. The aggregated information at the macroscopic and mesoscopic levels depends directly on the detailed information at the microscopic level.

In current times, computing capability and resources are ever increasing, along with the amount and quality of data available from actual railway planning and operations. For most of application contexts, it is necessary to model infrastructure networks at the microscopic level. For some detailed applications (e.g. calculation of running time and energy consumption), it is impractical or even impossible for the calculation to be carried out only at the macroscopic and mesoscopic levels. However, a high level of abstraction at the mesoscopic and macroscopic levels is useful to present and discover the attributes and relationships of infrastructure objects without overwhelming users with too much detail. Moreover, the abstracted level will provide a new perspective to observe the system performance. For example, the investigation at a mesoscopic level provides a moderate and balanced level of detail for identifying bottlenecks, although they are aggregated from the results calculated at microscopic level.

2.3 Modelling of Rolling Stocks

2.3.1 Applications and Level of Details

Modelling of rolling stocks for railway planning and operations includes the modelling of vehicles and the modelling of trains. A vehicle can be a locomotive, a railway wagon (without engine and motive power) or a multiple unit, whereas a train consists of one or several vehicles. The information of rolling stocks can be used for the following applications:

- Calculation of running time and energy consumption
- Design of timetables
- Resource planning for vehicles
- Maintenance

- Operations control and dispatching
- Capacity research

Depending on the application contexts, there are also several different levels of details in the modelling of rolling stocks. At the macroscopic level, a train can be modelled as a mass point, which is often used for running time calculations without high requirements of accuracy. Trains are usually represented at macroscopic level to hide detailed information. For example, a point can represent the head of a train in a distance-time diagram during timetabling.

At the mesoscopic level, a train is modelled as a homogeneous strip. Here, the attributes of running dynamics are assumed to be equally distributed within the homogenous strip. This strip can be used to represent the exact position of a train with consideration of the length of the train. For most applications for railway planning and operations, a model of rolling stocks at a mesoscopic level can provide a sufficient level of detail for the calculation of running time. In Figure 2-17, the trains are shown in the simulation tool PULSim at a mesoscopic level with red strips.

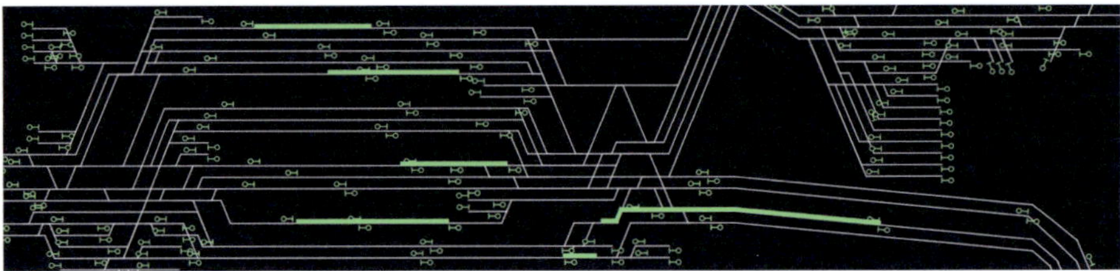

Figure 2-17: An example of trains at a mesoscopic level in the Software PULSim

To obtain the highest possible level of detail, all the components of rolling stocks are explicitly modelled at the microscopic level. A train will be modelled as a set of vehicles, and the position and characteristics of running dynamics for each vehicle will be investigated in detail. An example of a freight train at a microscopic level is shown in Figure 2-18, in which the train is running on a track with varied track gradients. In addition, the loading and the weight of each vehicle comprising the train vary. Here, all the vehicles are modelled separately and the attributes of running dynamics, e.g. the running resistance, will be calculated for each individual vehicle. The model at the microscopic level with a non-homogeneous strip provides the most realistic information, but demands considerably great computational efforts. For the purpose of railway planning and operations, system performance and required accuracy should be carefully balanced.

Figure 2-18: An example of trains at a microscopic level

During the modelling process, the information of rolling stocks for each vehicle will be initially assigned as attributes (see Section 2.3.2), and the formation of a train will be defined afterwards (see Section 2.3.3). With the detailed information of rolling stocks at the microscopic level, the macroscopic and mesoscopic models can be aggregated according to the concrete requirements and application contexts.

For a specific **vehicle definition** or a **definition of train formation**, many instances of vehicles/trains of the definitions can share the same information. It is common to create a vehicle definition and a definition of train formation as a basis, by which a concrete instance of vehicle/train is referenced. The redundancy of vehicle/train information for the vehicles/trains with the same class can thereby be avoided, and the consistency for the information of the vehicles/trains can be maintained efficiently. In Section 2.3, the modelling of rolling stocks focuses on the description of vehicle definitions and definitions of train formation.

2.3.2   Attributes of a Vehicle Definition

The attributes of a vehicle definition consist of both general and specific attributes. General attributes describe the information required for all application contexts, including the identity of the vehicle definition, the information of the manufacturer, the type of the vehicle (locomotive, railway wagon or multiple unit), as well as the basic physical attributes of the vehicle definition (e.g. the geometry, the gauge of the track and the weight). A set of specific attributes describes a certain configuration of the vehicle definition. The following specific attributes are often used (e.g. in railML) for railway planning and operations:

- Engine and braking system (for locomotives and multiple units)

- Train protection system
- Configuration of seats and standing room (for passenger wagons)
- Running dynamics

The listed attributes are selected according to the concerned components of the train; the attributes of the engine system are used for locomotives and multiple units, whereas the attributes of the braking system are used for all the vehicles and trains. The type of power (e.g. steam, diesel or electric), the parameters of the engine (e.g. power, voltage for electric type) and the type of the braking system should also be specified.

The supported train protection systems should be specified separately. For example, an Intercity Express (ICE) train may support both German PZB (an intermittent train protection system, in German: Punktförmige Zugbeeinflussung) and LZB (a continuous train protection system, in German: Linienzugbeeinflussung) systems. Meanwhile the compatible levels of European Train Control System (ETCS) of the train should also be input. The information will be used to simulate the train run under different types of train protection systems (see Section 4.2).

In order to calculate the capacity of the train and the loads, the configuration of seats and standing room for passenger trains is necessary. The expected load in the case of the maximum amount of allowed passengers should be taken into consideration for the calculation of the maximum axle loads. Similarly, the maximum allowed weight and volume of goods are required to be known in order to calculate the capacity of freight trains.

For calculation of running dynamics, the parameters for defining tractive forces and speeds are required for locomotives and multiple units. Sometimes the parameters are given in the form of force-speed pairs. An example of the input value for force-speed pairs for the locomotive class BR 146 in RailSys (RMCON 2010) is shown in Figure 2-19. The highlighted row shows a range of tractive force from 265.474 to 256.843 kN, which corresponds to the speed range from 57.2 to 71.5 km/h. For each pair, the force and the speed are specified in a certain range. Hence the tractive force can be determined by a given velocity and vice versa.

von V [km/h]	von V [mph]	von F [kN]	bis V [km/h]	bis V [mph]	bis F [kN]	Typ
0.000	0.000	300.108	1.000	0.621	299.507	ge
1.000	0.621	299.507	2.000	1.243	298.907	ge
2.000	1.243	298.907	3.000	1.864	298.307	ge
3.000	1.864	298.307	4.000	2.485	297.707	ge
4.000	2.485	297.707	5.000	3.107	297.107	ge
5.000	3.107	297.107	6.000	3.728	296.506	ge
6.000	3.728	296.506	7.000	4.350	295.906	ge
7.000	4.350	295.906	8.000	4.971	295.306	ge
8.000	4.971	295.306	9.000	5.592	294.706	ge
9.000	5.592	294.706	10.000	6.214	294.106	ge
10.000	6.214	294.106	11.000	6.835	293.506	ge
11.000	6.835	293.506	12.000	7.456	292.905	ge
12.000	7.456	292.905	13.000	8.078	292.305	ge
13.000	8.078	292.305	14.000	8.699	291.705	ge
14.000	8.699	291.705	15.000	9.320	291.105	ge
15.000	9.320	291.105	16.000	9.942	290.505	ge
16.000	9.942	290.505	17.000	10.563	289.905	ge
17.000	10.563	289.905	18.000	11.184	289.304	ge
18.000	11.184	289.304	19.000	11.806	288.704	ge
19.000	11.806	288.704	20.000	12.427	288.104	ge
20.000	12.427	288.104	21.000	13.049	287.504	ge
21.000	13.049	287.504	22.000	13.670	286.904	ge
22.000	13.670	286.904	23.000	14.291	286.303	ge
23.000	14.291	286.303	24.000	14.913	285.703	ge
24.000	14.913	285.703	25.000	15.534	285.103	ge
25.000	15.534	285.103	26.000	16.155	284.503	ge
26.000	16.155	284.503	27.000	16.777	283.903	ge
27.000	16.777	283.903	28.000	17.398	283.303	ge
28.000	17.398	283.303	29.000	18.019	282.702	ge
29.000	18.019	282.702	30.000	18.641	282.102	ge
30.000	18.641	282.102	31.000	19.262	281.502	ge
31.000	19.262	281.502	32.000	19.884	280.902	ge
32.000	19.884	280.902	33.000	20.505	280.302	ge
33.000	20.505	280.302	34.000	21.126	279.702	ge
34.000	21.126	279.702	35.000	21.748	279.101	ge
35.000	21.748	279.101	36.000	22.369	278.501	ge
36.000	22.369	278.501	37.000	22.990	277.901	ge
37.000	22.990	277.901	38.000	23.612	277.301	ge
38.000	23.612	277.301	39.000	24.233	276.701	ge

Figure 2-19: Tractive Force and Speed Diagram for the BR 146 in RailSys

Similar to the parameters for defining tractive forces and speed, the parameters of braking forces and speeds can also be displayed in the form of force-speed pairs.

In addition to the tractive force and the braking forces, train resistance should also be considered for the calculation of running dynamics. The train resistance consists of at least running resistance, line resistance and starting resistance. The constant, linear and quadratic coefficients required for the calculation of running resistance are determined through the following equation:

$$R = a + b \cdot V + c \cdot V^2 \qquad\qquad (2\text{-}1)$$

Notations used:

R Running resistance [N]

a Constant coefficient [N]

b Linear coefficient [N·h/km]

c Quadratic coefficient [N·h^2/km^2]

V Speed of the train [km/h]

Line resistance is infrastructure-related, and can be calculated from the geometry information in the given infrastructure model (see Section 2.2.4). The starting resistance only presents an affect at the starting time, and is the determining factor in whether a locomotive is able to start.

2.3.3 Definition of train formation

A train can consist of locomotives, wagons or multiple units. After the formation of a considered train is defined, the attributes of said train can be aggregated from the included vehicle definitions.

A common method for carrying out train modelling is to build a train as a "container", in which the vehicle definition of each individual vehicle is established and maintained. This approach is very suitable for modelling freight trains, since the formation of the train is highly dynamic and frequently changes, often during the course of a train run during operations. It is necessary to specify the order of the vehicles and the supported coupling type, as well as the characteristics that are used to calculate the compatibility and the processing time for simulating shunting processes and operations of coupling/decoupling at microscopic levels.

Special attention should be paid to passenger trains, whose formation is often fixed. In the case of multiple units, it is possible to model a multiple unit either as a single vehicle, or as several vehicles. For the latter case, an individual vehicle is specified through a vehicle definition, and a multiple unit will be modelled in the form of a definition of train formation.

The general term "**unit definition**" is used as the super class of vehicle definition and definition of train formation. The unit definition only provides the basic attributes shared by vehicle definitions and definitions of train formation (e. g. the length and the total weight).

Within a definition of train formation, a list of unit definitions is maintained to describe the formation of a train. Since the unit definition is the super class of vehicle definitions and definitions of train formation, it is possible that several definitions of train formation are combined. In this case, each individual definition of train formation represents a multiple unit. The class diagram of a unit definition, vehicle definition and definition of train formation is shown in Figure 2-20.

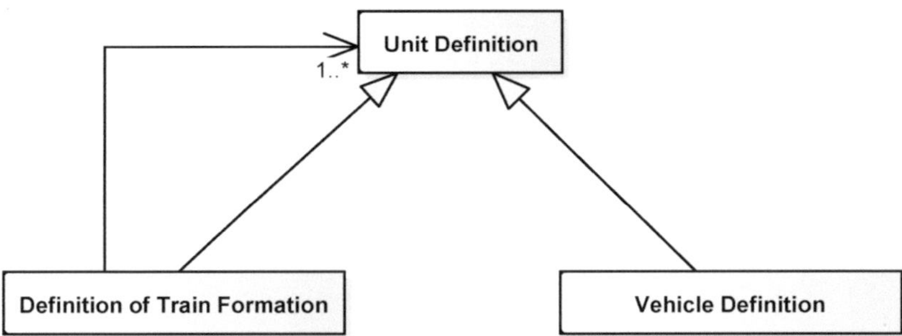

Figure 2-20: Class diagram of a unit definition, vehicle definition and definition of train formation

When considering passenger transport, the establishment of a fixed definition of train formation is usual. For a train that consists of several multiple units, the flexibility exists to build the train as a combination from the definitions of train formation for the multiple units. During simulation, each multiple unit (with a certain definition of train formation) can be either operated individually or be combined together as a whole in a flexible manner. A complete train can be composed of two multiple units (sub-trains), especially for so-called "coupling and sharing" operations, which are rapidly gaining importance in recent times. Depending on the operational conditions, the sub-trains can either share the same definition of train formation or use different definitions of train formation. In a coupling and sharing operation, one train can be separated into two individual sub-trains at a given intermediate station, with the two sub-trains running towards different destinations. After the two sub-trains return, they will be coupled again at a station as one train. In this way, the resources are utilised efficiently according to the traffic demand. A "complete" definition of train formation that combines the two sub-trains is defined before separation and after coupling. If the two sub-trains share the same formation, each of them will share the same definition of train formation as that of the multiple-unit. Otherwise, the definition of train formation for each sub-train will be defined separately. To successfully accomplish a coupling and sharing operation, the signalling systems for shunting movements and the shunting rules of infrastructure occupancy should be integrated. A detailed explanation of the coupling and sharing operation will be presented in Section 2.4.3.

If a passenger train consists of several fixed wagons with different classes of locomotives, it is practical to define a fixed definition of train formation without locomotives. In this case, different classes of locomotives can be combined with the definition of train formation (without locomotives) to build a new flexible definition with a specific locomotive. An example is shown in Figure 2-21, in which the definition of train formation "td" includes all the wagons, without the consideration of locomotives. It can be combined with a suitable class of locomotive (e. g. BR 146) to form a complete train, which is specified by the definition of train formation "rt".

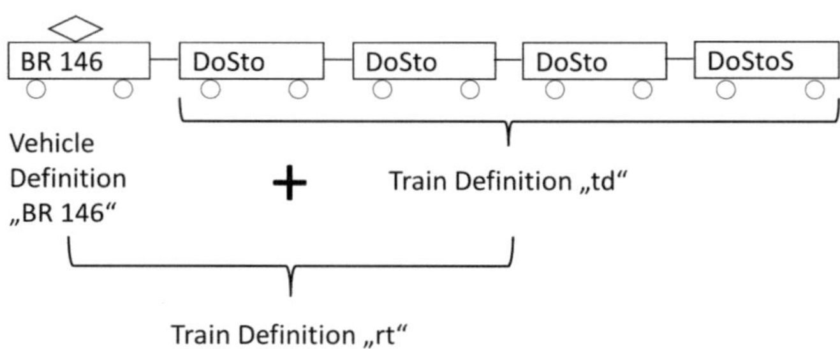

Figure 2-21: An example of train formation

For mesoscopic and macroscopic models, the definition of a train will be simplified to a level of only one "abstract" single vehicle, which aggregates the information of the entire train. The attributes of the abstract single vehicle are derived and assigned for the complete train.

2.4 Modelling of Railway Operations

2.4.1 Multi-Scale Models of Railway Operations

Railway operations cover a wide-ranging spectrum of activities, including the planning of railway operations and railway operations control. For long term planning over a large-scale network, macroscopic models are often used, because detailed information and attributes of railway operations are either unavailable or unable to be handled. At the macroscopic level, the model of railway operations is based on the macroscopic model of infrastructure and rolling stocks.

For railway operations control and detailed planning and calculation (e.g. timetabling and running time calculation), railway operations should be modelled at a microscopic level. The detailed attributes of infrastructure layout, geometry, signalling, as well as rolling stocks are required to ensure conflict-free operations. A microscopic model of operations is always based on a microscopic infrastructure model. However, to reduce required computational efforts, it is common that a train is modelled as a mass point on a macroscopic level or a homogenous strip on a mesoscopic level. The macroscopic or mesoscopic model of rolling stocks can also provide accurate results close to reality (Brünger, Dahlhaus 2014).

In a multi-scale framework, the macroscopic and microscopic models of operations can work together for railway planning and operations. A two-step approach is often used in timetabling (Kroon et al. 2014). This process consists firstly of the scheduling of the arrival/departure/passing time of trains at macroscopic nodes (stations and junctions), without considering the detailed block sections and paths. After the timetable has been optimised at the macroscopic level, microscopic timetabling is carried out in order to derive a conflict-free timetable. Similarly, for railway operations and train dispatching, the multi-scale framework can be applied to avoid high computational complexity (Cui 2010). A basic dispatching solution is established and optimised at the macroscopic level, and the solution will be further elaborated at the microscopic level to generate a final, detailed dispatching timetable.

The simulation output of railway operations at the microscopic level can also be used to evaluate the quality of railway services at mesoscopic or macroscopic levels. For example, the line headway at the macroscopic level can be computed after the detailed running time calculation at each block section. In (Martin et al. 2014), the quality of service and bottlenecks are analysed based on the basic structure (see Section 2.2.6) at the mesoscopic level. The occupation time and hindrance time of each basic structure are calculated from the recorded log files generated from microscopic simulation.

Most of the applications of simulation are carried out at either microscopic or macroscopic levels. The model of railway operations at the mesoscopic level is used to analyse the results with a moderate level of detail. Depending on the purpose of research, there are diverse designs and attributes for mesoscopic models. Therefore, the scope of this book will focus only on the models of railway operations at the macroscopic level and microscopic levels.

Modelling of the operating program and timetable form the basis for the model of railway operations (see Sections 2.4.2 and 2.4.3). Thereafter, the rolling stock assignment and crew schedule can be designed (see Section 2.4.4). The operations control, which establishes the coordination between all the components of a railway system, will be discussed in Section 2.4.5.

2.4.2 Model of Rough Operating Programs and Timetables at a Macroscopic Level

An **operating program** (Martin 2015) is comprised of all the details of railway operations, which justify the construction of railway infrastructure and determine infrastructure layout. A complete operating program covers the data-related description of the system performance and requirements of quality. Among other factors, it consists of:

- Amount of train runs
- Structure, sequence, and train characteristics of the train runs, as well as the relationship influenced by the design of the infrastructure and the system performance
- Temporal allocation of the train runs

A rough operating program can be modelled for railway operations at a macroscopic level. It describes the basic concepts of the supply-oriented railway operations for long term planning, including train characteristics, the structure of train runs, as well as the sequence of stations along the train runs at a macroscopic level. Train characteristics describe the type of the train (e.g. passenger trains or freight trains) and the general attributes of the rolling stocks (e.g. electric-hauled trains or diesel-hauled trains). The structure of train runs defines the proportion of each train group, which is established through the grouping of those train runs with the same attributes, train characteristics and stations. With the defined structure of train runs, the amount of train runs can be a variable, which is often used to determine the capacity of an investigated network. In a rough operating program, the detailed paths and the exact arrival/departure/passing times of a train run at a microscopic level are not relevant. Furthermore, the connection between two train runs should be designed to ensure the desired transfer. During a transfer, the feeder train brings passengers/goods to the connecting train. The transfer can also be a two-way feeder service, which means that a train in a transfer can be both the feeder train and the connecting train. In case one of the trains is delayed, special rules and thresholds are necessary to model and simulate the behaviour of the feeder train and the connecting train. The design of the rules to identify connection conflicts is discussed in Section 5.1.2. An overview of the model of rough operating programs is shown in Figure 2-22:

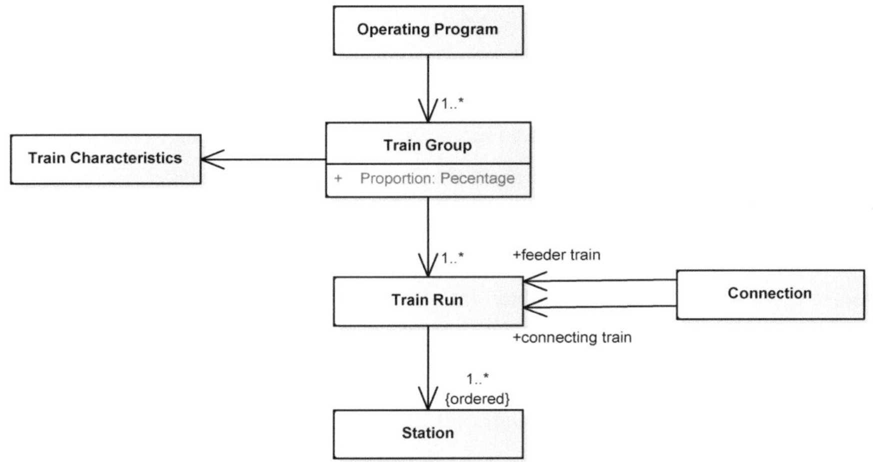

Figure 2-22: Model of rough operating programs

For a rough operating program, timetables can be further developed at a macroscopic level. A timetable defines a number of **train runs**, whereby each train run has a fixed train number as an identity during operations. In macroscopic timetables, only the arrival/departure/passing time at each station is defined, and is stored in an **entry**. A timetable consists of a series of entries sorted by the scheduled arrival/departure/passing times. Timetables can be categorised as random timetables, non-cyclic timetables, partially cyclic timetables and cyclic timetables (Martin 2016). Within partially cyclic and cyclic timetables, trains operating at a fixed frequency are organised in a train group.

The running time between two stations is roughly calculated for each day type (e.g. normal working day, weekend, or holiday) and time group (e.g. peak hour or non-peak hour), with consideration of the minimum headway between two trains on a line. The recovery time and the buffer time should be introduced as well to ensure stable and robust operations for the given timetables. For cyclic timetabling, the frequency of the services is also defined, which corresponds to the rough operating program and the minimum headway.

Mathematic optimisation models can be used to construct timetables and to improve the robustness of services at a macroscopic level. For example, the model to solve the Periodic Event Scheduling Problem (PESP) was developed in (Serafini, Ukovich 1989) for scheduling cyclic timetables. In (Caprara et al. 2002), the arc-based timetabling model was designed for non-cyclic timetabling. An overview of macroscopic timetabling and optimisation is given in (Kroon et al. 2014).

Timetables at macroscopic levels are the basis for rolling stock assignment and crew scheduling (see Section 2.4.4). The whole procedure for timetabling, planning of rolling stock assignments and crew scheduling can be executed iteratively. The macroscopic timetables are also used to exchange information among different applications and operators. For example, the Association of German Transport Companies (VDV) specified the data model of timetables for urban public transport as an open interface (VDV 2013). The defined macroscopic timetables and operating programs can be shared among different transport authorities and operators for public transport planning, operations control, and passenger information.

Due to the high requirements of accuracy for railway planning and operations, a detailed timetable and paths for train runs is necessary. Therefore, the rough operating programs and macroscopic timetables should be further examined and elaborated at a microscopic level.

2.4.3 Model of Microscopic Timetables

At a microscopic level, it is required to construct a conflict-free timetable according to the detailed information of block sections, paths, signalling systems, and rolling stocks, along with specifying the exact path of a train within it. A microscopic timetable should be constructed with consideration of all the required time-related components, including the regular transport time, the scheduled waiting time and the scheduled synchronisation time.

The model of microscopic timetables and paths is based on the microscopic infrastructure model. Similar to the macroscopic timetables, a series of entries are included in a microscopic timetable. The term "**reference point**" of a timetable is defined as a general term for a stop or a measurement point, and each entry is referenced by a reference point of running time. Not only the scheduled arrival/departure/passing time of each stop, but also the microscopic path along the entire train run should be specified. The block sections of the signalling system will be derived from the microscopic paths afterwards. It is very common to model microscopic paths referenced by each station, limited by a starting point and an ending point at station boundaries, in which case the microscopic path at each station is referred to as a station route. At a station, it is possible that several station routes have the same starting and ending point. During simulation/operations, the train will take the pre-defined station route at a station. The design of station routes results from the structure of the railway operations control. In (UIC 2013), the term station is defined as "points of a network where overtaking, crossing or direction reversals are possible". The whole network is divided into several stations as control territories, which are connected by open tracks. The station routes inside a station specify the entry and the exit point at station boundaries. Hence the complete path can be determined via connecting the station routes along the whole train run. The model based on station routes is applied widely for railway planning and operations (e.g.: (RMCON 2010), (VIA Con 2011)).

Another method to model the microscopic path is to define the section between two reference points (stops or measurement points of station). This concept is usually used for urban public transport, for which the concept of a station refers to the stopping places for scheduled stops.

Coupling and sharing operations allow passenger trains to be decoupled flexibly into two sub-trains. In this case, several **Parts of Train Run (PTR)** should be defined under a train run. Each part of train run is operated along a specific path with a sub-train number, and the definition of train formation for each part of a train run should be specified. The example described in Section 2.3.3 is shown in Figure 2-23. Four parts of a train run (PTR 0, 1, 2, and 3) are defined for two sub-trains (sub-train 1 and sub-train 2), as well as for two complete trains (complete train 0 and complete train 1) before separation and after coupling. The two sub-trains share the same definition of train formation ST, and the two complete trains share the same definition of train formation CT. The four parts of train run refer to two different definitions of train formation (ST and CT), respectively.

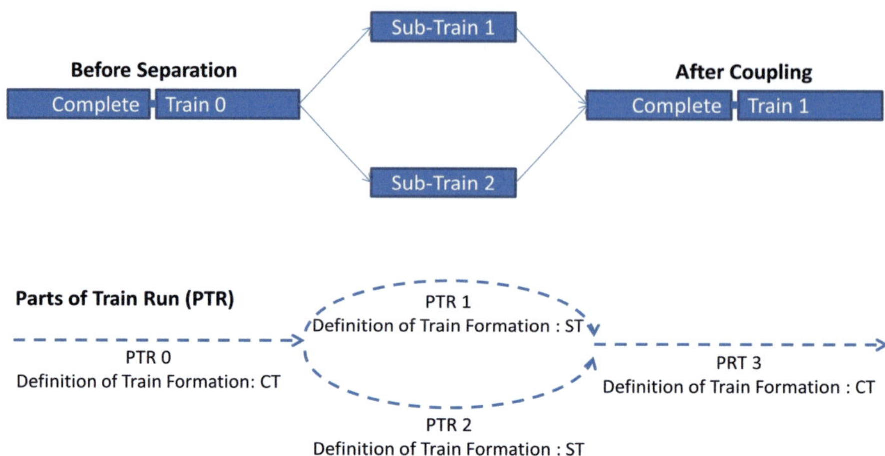

Figure 2-23: Example of parts of train run for coupling and sharing operations

The class diagram of train runs based on station routes for microscopic timetables is shown in Figure 2-24. The detailed definition of train formation (see Section 2.3.3) should be specified for each train run or part of train run. Hence the running dynamic characteristics of the train, the supported train protection system, as well as the electric system (for electric-hauled trains) are known. Combined with the geometry information and block sections defined in the microscopic path, the exact running time can be calculated, and the behaviours of train runs can be simulated at a microscopic level.

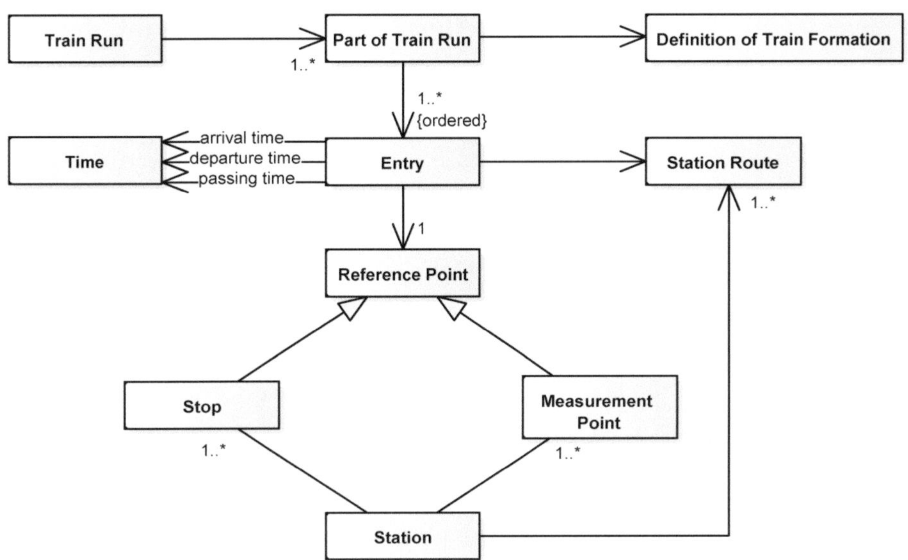

Figure 2-24: Microscopic timetable model

2.4.4   Model of Rolling Stock Assignment and Crew Scheduling

After the construction of the timetables, the scheduling (also called rostering) of rolling stocks and crews is necessary for railway operations. A day-to-day operation is scheduled primarily, and a rescheduling process is carried out by dispatchers during daily operations control.

This section focuses on the model of rolling stock assignment and crew scheduling for passenger transport. The data of infrastructure, rolling stocks and timetables is used as the basis to model rolling stock assignment and crew scheduling. Not only the stations and the open tracks along a train run, but also the depots as well as the lines between the start/end stations of the train run to the depots should be given. Depots are treated similarly to stations. The same modelling method for infrastructure can be applied for the extended network with depots. For timetable models, the vehicle definition and the definition of train formation are used for a part of train run (see Section 2.4.3), while the concrete vehicles and trains as instances of the definitions are used to model rolling stock assignments and crew scheduling. The difference between vehicles/trains and the definitions can be found in Section 2.3.1.

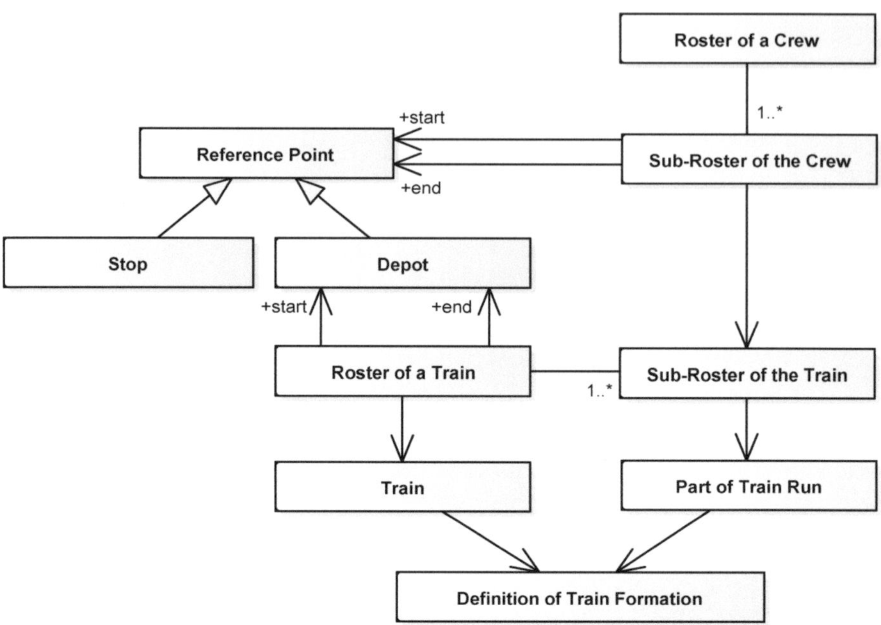

Figure 2-25: Roster of a train and roster of a vehicle for passenger transport

The **roster of a train** and the **roster of a crew** are the basic items in the model of rolling stock assignment and crew scheduling (see Figure 2-25). A roster of a train models the usage of the train during a day-to-day operation, and begins and ends at a depot. The starting depot and the ending depot of a train may vary. In one day, a given train is usually under operation for several different parts of train run (see Section 2.4.3). Therefore, a roster of a train consists of several sub-rosters of the train. For each sub-roster of the train, a part of train run is assigned, which is bound to the same definition of train formation of the concerned train.

For each staff member, a crew roster is created, which includes several sub-rosters of the crew. One or more sub-rosters of the crew are assigned for a sub-roster of a train. A sub-roster of crews can start or end at a depot or a stop. Additional information of crews, including the regulations of working hours and pauses, the salary rates of crews, the planned vacations, as well as the preferences of crews, should be given as constraints for crew scheduling.

For freight transport, depots are used for locomotives. The freight wagons are traced and monitored for both regular movements and shunting movements. The management of freight wagons and staff in freight transport is more flexible than in the case of passenger transport. Depending on the specific operations to be carried out, several constraints used in the assignment of rolling stocks and crews should be defined. For example, in order to ensure that the maximum amount of allowed working hours is not exceeded, specific lines for a given locomotive driver as well as stops for interchanging locomotive drivers should be defined. Sometimes a locomotive is bound to a specific driver inside a certain area. The management of freight wagons and staff should be modelled based on the relevant constraints.

Based on the roster of trains and crews, the assignment of rolling stocks and crews can be scheduled. Mathematic models and simulation methods can be applied to:

- Minimise the requirements of rolling stocks
- Minimise empty mileage (e.g. train runs from depot to station, or from station to station without passengers and goods)
- Minimise non-productive downtimes
- Minimise the costs of crews.

The approaches to plan and to optimise the usage of rolling stocks and crew scheduling are presented in (Bussieck et al. 1997), (Ghoseiri et al. 2004), (Fioole et al. 2006), (Ernst et al. 2001) and (Voss, Daduna 2001). The data model of rolling stock assignment and crew scheduling should be independent of the applied algorithm.

2.4.5 Model of Operations Control

Operations control can be perceived as the process of managing the interaction between infrastructure, rolling stocks and operating programs. It also includes the process of short-term planning against unscheduled disturbances during daily railway operations. The same models of timetables, rolling stock assignment and crew scheduling are utilised for both the planning of railway operations and traffic control. In addition, the aspects for monitoring and dispatching systems should be considered. One of the most important functions of operations control is to allocate infrastructure resources. An **infrastructure resource** was originally defined in (Cui 2010) as the minimum set of infrastructure in railway operations, which is requested, allocated, and released as a basic unit at the same time.

The definition of the infrastructure resource depends on the application context. In a signalised network with fixed block section, an infrastructure resource is a block section or a path component (if partial route releasing is applied). For a signalling system containing overlaps, an overlap will also be defined as a path component (infrastructure resource). Depending on the actual operational situation, variable overlaps can be applied in order to adapt to the change of train velocity. Hence, the infrastructure resources based on overlaps of the block section are variable accordingly.

For railway operations with moving blocks or in a non-signalised network, an infrastructure resource can be defined based on the directed edges of non-junction-type infrastructure elements (tracks) or junction-type elements (e.g. turnouts, crossings).

For the monitoring system, it is required to identify and report the position of trains in real time. The information used for positioning a train and measuring the actual running time of the train consists of:

- Position of reference points in the infrastructure network
- Position of the train (the head of the train and/or the rear of the train)
- Measurement time

Depending on the signalling system, the position can be reported to the Operations Control Centre (OCC) from either infrastructure side devices and/or trainborne devices. Further actions will be trigged once a train arrives or leaves a certain position. For example, the block section should be assigned or released once a train approaches or leaves a block section. It is also necessary to update the actual arrival/departure/passing time of a train at stops and measurement points. The actual time will be compared with the scheduled arrival/departure/passing time in order to calculate delays and possible conflicts. If conflicts have been identified for the actual operational situation, a dispatching system will be activated to resolve potential conflicts.

Knowledge regarding the state of infrastructure is required in order for a simulation tool to assign infrastructure resources and to guide train movements. For example, the information related to the vacancy or occupancy of block sections is used to examine potential conflicts or deadlocks. The aspect of a signal and the position of switchblades are important in simulating the behaviours of interlocking systems.

The monitoring system also should collect or generate related to disturbances, which are the sources of delays. They are modelled as the additional time extensions of train runs due to random events, e.g. technical failures on railway infrastructure and rolling stocks, accidents, stochastic driving styles or variations of passenger boarding/alighting time. It is referenced to a position (e.g. at a stop or at an open track section between two stations) and a certain train group/train type. Disturbances are used to simulate the stochastic influences during operational simulation. In Section 6.4, the generation of disturbances and calibration of the disturbance parameters will be discussed in detail.

For the dispatching system, it is also required to register the duration, the reason, as well as the predicted impacts of a disturbance, so that a suitable dispatching measure can be taken. The statistical distribution of the disturbances, the impacts of disturbances, as well as the relationship between the delays and the reasons of delays, can be further evaluated and analysed by evaluation packages.

Depending on the specific (automatic) dispatching solution, the applied model of a dispatching system in simulation software may be varied. The dispatching measures can be categorised as time-related and location-related dispatching (Martin 1995), (Cui 2010). With time-related dispatching, only the velocity of trains, the waiting time, and the resulting transport time are changed. The sequences of trains may be changed as well through setting waiting times for the conflicting trains. More complicated dispatching measures are applied with location-related dispatching, which includes overtaking, detouring, shortening scheduled path, cancelling of train runs, reassignment of rolling stocks, and rescheduling of crews. There is no standard way to model dispatching measures due to the diversity of available dispatching algorithms. However, the resulting dispatching timetable can be modelled as a "normal" timetable. The dispatching timetable can also be generated with different level of details. In (Cui 2010), a multi-scale dispatching method is proposed. With multi-scale dispatching, a macroscopic dispatching timetable can firstly be generated and optimised. Based on the macroscopic dispatching timetable, a microscopic timetable will be further elaborated, which will provide the detailed information of the rescheduled train runs.

After the structural perspectives of railway system are modelled, the components of the system will communicate and interact with each other. The behavioural perspectives and the workflow of railway simulation will be discussed in Chapter 3.

3 Building the Workflow of Railway Simulation – Behavioural Perspective

3.1 Overview of the Workflow in Railway Simulation

The workflow of railway simulation describes the behaviour of the modelled railway system and the interaction among its components. The model of operations specifies the schedule and the organisation of train runs. The interaction between infrastructure and rolling stocks determines how infrastructure resources are requested and allocated. All of the behaviour and the interactions of a simulation model are implemented through control flows and message flows, which are driven by the applied workflow of a simulation tool.

The workflow of a simulation model can be categorised from three different dimensions (Law 2007):

Static vs. Dynamic simulation: A static simulation can be used to represent a system at one particular time, or a system that is does not evolve over time, while a dynamic simulation is a representation of a system evolving over time.
In railway simulation, there are two types of simulation: **synchronous simulation** and **asynchronous simulation**. Synchronous simulation is a dynamic simulation, in which all train runs are simulated simultaneously along the passage of time. It represents the behaviour of a railway system and the interaction among its components realistically. The conflicts and the resulting waiting time can be simulated identically to reality. In asynchronous simulation, trains are introduced into the simulation system according to their pre-defined priorities, respectively. The train introduced firstly has a higher priority and should not be influenced by lower priority trains. The lower priority trains are inserted to the system after the train runs of higher priority trains have been fixed. Since the blocking time of all individual trains has been calculated in advance, asynchronous simulation can be categorised as static simulation. The process of introducing trains in asynchronous simulation does not depend on the passage of time.
Compared with asynchronous simulation, synchronous simulation is more practical to simulate the interaction and the conflicts among trains during railway operations. It is

also very popular to use synchronous simulation at a microscopic level to study the occurrence of delays and delay propagation. Asynchronous simulation concerns the overall usage of infrastructure resources and capacity. It is often used for scheduling, where the train runs are planned and handled according to the priority of trains. Therefore, synchronous simulation is suitable for a purely operational simulation, and asynchronous simulation is often used for a purely timetable-based simulation (see Section 3.3).

In further research, the principle of asynchronous simulation could be implemented in a synchronous simulation. For example, if additional freight trains are to be introduced without influencing existing train runs, the principle of asynchronous simulation could be applied. The existing train runs are initially introduced with high priorities. The freight trains can afterwards be simulated synchronously or asynchronously in order to determine the available train paths. Meanwhile, a synchronous simulation can be carried out for a certain train group in an asynchronous simulation.

In the case of both synchronous and asynchronous simulations, the blocking time of train runs should be established. The calculation of blocking time for an individual train run in asynchronous simulation can be perceived as a synchronous simulation, in which only one train is simulated with the passage of time. In this book, the workflow of synchronous simulation will be presented in detail. In Section 3.2, the framework of discrete dynamic simulation (synchronous simulation) is designed. The simulation of an individual train run and the approaches to resolve conflicts among several trains will be presented in Chapter 4 and Chapter 5. A well-known issue of synchronous is the deadlock problem, the details and solution of which are explained in Section 5.2.

Continuous vs. Discrete simulation: Both continuous simulation and discrete simulation are forms of dynamic simulation, in which the passage of time is concerned. In a continuous simulation, the state variables that represent the status of components change continuously with respect to time. An example of a continuous simulation is the model of a running time calculation, in which the movement of a train is continuously varying. Differential equations can be used to derive the value of the state variables over time analytically. However, if the differential equations are not available or too complicated, the continuous simulation has to be solved with a numerical approach, which is implemented through discrete simulation.

During discrete simulation, the state variables are determined and changed at certain points in time. The time interval between two points of time can be fixed (in time-driven simulation) or variable (event-driving simulation). The design of discrete simulations will be discussed in Section 3.2.

In railway simulation, the combined continuous-discrete simulation is applied, since a railway system cannot be simply defined as a completely continuous or discrete system. One reason for that is for example, the continuous changing of train movement over time. However, the system may be influenced by certain discrete events (e.g. state changing of the signalling system). The continuous changing of train movement can also trigger a discrete event (e.g. releasing a block section when the rear of a train leaves a signal releasing point). Indeed, the continuously changed train movements are often simulated through discrete simulation. The running dynamics of a train are calculated by interpolating train movements at the time or at the positon between two discrete time points. In this book, the framework of discrete simulation will be discussed in Section 3.2, and the simulation for running time calculation will be established explained in Section 4.1.

Deterministic vs. Stochastic simulation: In deterministic simulation, the input values and parameters are fixed, which leads to a "determined" output without probabilistic components. In railway simulation, deterministic simulation is often used for construction of timetables, and is therefore also known as timetable simulation. In stochastic simulation, some input values and parameters are modelled stochastically. The system behaviour can be investigated with consideration of random influences and variations. In railway simulation, stochastic simulation is used to determine the robustness and stability of an operating program, as well as the quality of service in railway operations. Therefore, it is also referred to as operational simulation. In railway operations, stochastic influences should also be considered for train dispatching, in order to predict and adapt to the frequently changing operational situation. Otherwise, the robustness of the derived dispatching timetable will be limited. Since multiple simulations for many sets of stochastic inputs will be carried out in stochastic simulation, the term "multiple simulations" is sometimes used for stochastic simulation. Correspondingly, the term "single simulation" is used to refer to deterministic simulation. In this book, the terms "timetable simulation" and "operational simulation" are used for deterministic simulation and operational simulation, respectively. The workflows of timetable simulation and operational simulation are discussed in Section 3.3, and the methods to calibrate the parameters and inputs are presented in Chapter 6.

3.2 Discrete (Dynamic) Simulation

3.2.1 Time-driven Simulation

In **time-driven simulation**, all of the train runs and the interaction between trains and infrastructure are simulated synchronously along a series of time points. The time interval between two time points is fixed. Since time-driven simulation is a type of dynamic simulation, a system clock is maintained to control the passage of time. During a time-driven simulation, the time of the system clock is increased by a fixed time interval. The train runs and infrastructure will be simulated at each discrete time point accordingly.

In a single processing step of a discrete time point, the position of each train will be forecasted initially. A running time calculation is carried out to determine the changing of the position and the speed of trains during the last time interval. Once a train requests to occupy a new infrastructure resource at that time interval, the system will examine if the requested infrastructure resource is possible to be allocated. It is necessary to ensure conflict-free and deadlock-free situations during the allocation of infrastructure resources (see Chapter 5). If the situation is not suitable for a train to occupy the requested infrastructure resource (due to conflicts or deadlocks), the request is pended. After the requested resources are allocated or pended, the position and the velocity of the train at the next time point can be determined. The states of trains and infrastructure will be updated for that time point. The workflow of a time-driven simulation is shown in Figure 3-1.

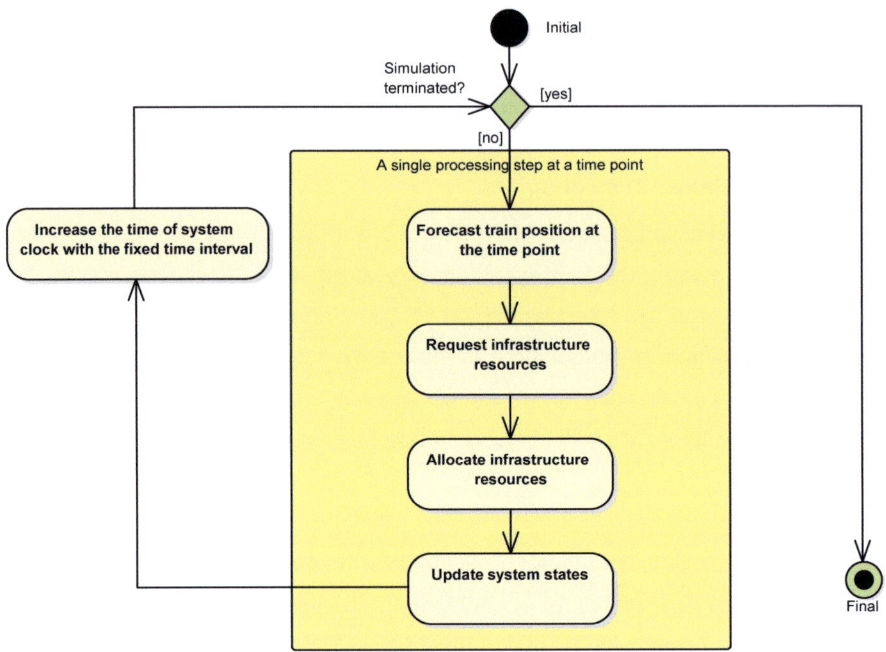

Figure 3-1: Workflow of time-driven simulation

The workflow of time-driven simulation is simple to implement. It is suitable for the simulation of railway operations with simplified interactions between trains and infrastructure. However, the disadvantages concerning the accuracy and the efficiency of time-driven simulation should also be considered. The events of train runs do not always take place exactly at each fixed time point, in which case time-driven simulation will limit the accuracy of the simulation. During a time-driven simulation, the change of system status at many fixed time points may be insignificant or even irrelevant. The unnecessary computational efforts due to calculations at these time points will therefore decrease system performance and efficiency.

For example, if a train is supposed to occupy an infrastructure resource at a time point t between the time interval from time T_1 to T_2 (T_1 and T_2 are fixed time points), the train can only occupy the requested infrastructure resource until the time point T_2. The error of the starting occupancy time in the time-driven simulation is $T_2 - t.$, which leads to an unreliable prediction of conflicts for train dispatching.

Another example related to system performance and efficiency can be seen during a running time calculation. If a train is running along a long track with constant speed during its cruising phase, the state of the train is updated regularly in the time-driven simulation. However, the computational efforts required for updating the train state are not necessary, since the state of the train can be easily calculated until the end of the cruising phase.

It is problematic to set the length of the time interval in time-driven simulation appropriately. An excessively large time interval can result in a high number of errors due to the difference between the exact event time and the fixed time point. An insufficient time interval will lead to high computational efforts. Furthermore, the results of time-driven simulation are not identical with differing lengths of time intervals. Therefore, event-driven simulation is more suitable for railway synchronous simulation with high accuracy and efficiency.

3.2.2 Event-driven Simulation

In **event-driven simulation**, the system evolves over a series of time points. At minimum, one certain event occurs at each time point. While the time interval between two time points is fixed in time-driven simulation, the events can be scheduled realistically according to the exact occurrence time in event-driven simulation.

An **event** is defined as an occurrence that may change the state of the system at a certain time point. There are many different implementations of event-driven simulation in railway simulation. In the rest of this section, the workflow of event-driven simulation is illustrated based on its implementation in the software PULSim. Although the attributes of the involved components (infrastructure, rolling stocks, and operations) may be changed continuously during a simulation process, only the specific occurrence of the changing of state, which causes interactions between two or more components or triggers another event, is considered a discrete event. Otherwise, the changes of the states, e.g. the continuous changing of train velocity, will be handled at each discrete time point by the train itself, without requiring an explicit event (see Section 4.1).

With the design of event-driven simulation applied in PULSim, the changes of system states, which are irrelevant to other components and events, can be processed and encapsulated inside the involved component, and the related computational efforts can be allocated to the individual component. Therefore, complexity of event-driven systems will be reduced with high scalability.

In particular, the isolation of running time calculations for an individual train provides the possibility to schedule events dynamically with low computational efforts. The special design applied in the software PULSim can achieve a high system performance without sacrificing accuracy. In PULSim, the events for a train will be scheduled at the time when a movement authority is granted. A **movement authority** is "the permission given to a train to move up to a defined position" (Anders et al. 2009). For example, if a train is allowed to occupy a requested block section, a movement authority for the train to move up to the ending main signal of the requested block section will be granted. Once a new movement authority is granted to the train, it will recalculate the train movements (running time, position, and velocity) until the **End Of Authority (EOA)**. The EOA is the location to which the train is authorised to proceed and at which the target speed is zero (Ligier, Guido 2012).

For a granted movement authority, a train is allowed to continue its train run until the EOA. Therefore, the running dynamics of the train until the EOA can be calculated once, and a batch of relevant events can be scheduled at one time for a granted movement authority. An illustration of the process is shown in Figure 3-2. At the point A, a movement authority with EOA at the main signal S2 is granted. For the section from A to S2, the speed profile is calculated and the events are scheduled. At the point B, a new movement authority with EOA at the main signal S3 is granted. The events for the section from B to S3 are scheduled. Meanwhile, the events scheduled between B and S2 will be discarded, since the running dynamics from B to S2 will be updated with the newly calculated speed profile. In case the train has not received the movement authority at the time at which it reaches at the main signal S2, the originally scheduled events from A to S2 are still in effect. Similarly, further events can be scheduled in a batch, based on the granted movement authorities along the path of the train.

It is possible that a new event can be triggered by a scheduled event. Since the occurrence of the new event is still based on the actual operational situation, the calculated speed profile can be reused to schedule the new event. Afterwards, the new event will be inserted into the event queue.

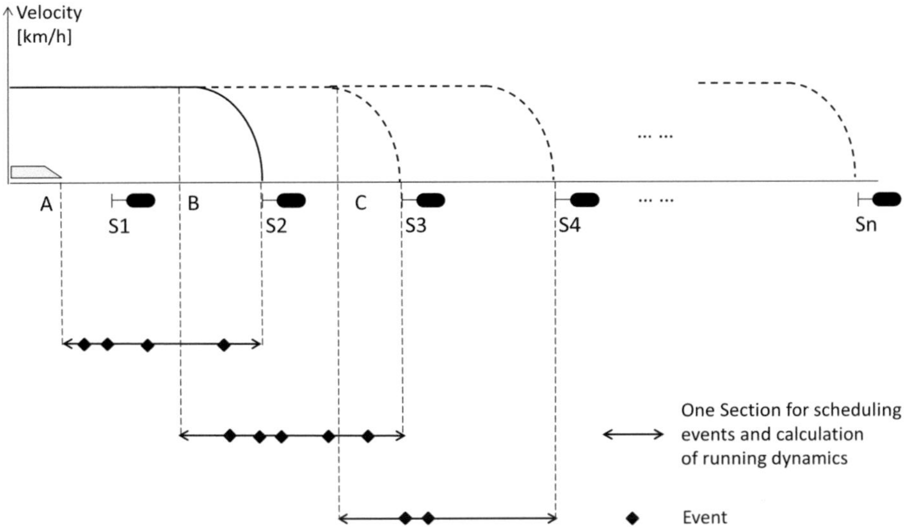

Figure 3-2: Event scheduling and calculation of running dynamics

In a conventional event-driven simulation system, a new event is usually scheduled after the occurrence of an event (Law 2007), which results in high computational efforts for a running time calculation in railway simulation. If each event is scheduled one by one, the running dynamics for each event still have to be calculated until the EOA, in order to derive the speed profile. The mechanism for event scheduling based on the granted movement authorities can avoid this repeated computation. All the events until the EOA can be scheduled in a batch through sharing the same speed profile. In addition, the system complexity is reduced since the interaction between the involved components and the change of running dynamics is isolated.

In the software PULSim, the important events, which may trigger new events and/or lead to interactions among components, are defined as the following:

- "Train start" event: a train is scheduled to start its train run. The state of the train will be initialised, and an event to request an infrastructure resource will be triggered.
- "Train terminate" event: a train is scheduled to terminate its train run. An event to release infrastructure resources will be triggered.

- "Train update location" event: this event is used to update train position, in the case that a train arrives at a certain point (e.g. a stop or a signal), at which another event will be raised.

- "Train arrive/departure operational point" event: the actual arrive/departure/passing time at an operational point (a stop or a measurement point for running time) will be recorded and compared. The actual delay of the train is calculated, which will be used for predicting the train run and train dispatching.

- "Train arrive request point" event: if a train approaches a certain point (e.g. distant signal), it requests to occupy the next block section through raising a "request infrastructure resource" event. The train speed will be regulated according to whether or not the next block section is allowed to be occupied.

- "Train arrive main signal" event: if a train arrives at a main signal, at which the next block section has not yet been granted, the train should stop at the signal and release the overlap after the main signal. If the main signal is a combination signal, a "request infrastructure resource" event will be triggered.

- "Grant movement authority" event: If a grant movement authority event is raised, the events of the train run should be rescheduled. The time and the train position to trigger events until the EOA for the granted movement authority will be searched for, and the events will be scheduled accordingly.

- "Train leave signal/route releasing point" event: this event will trigger an action to release the infrastructure resource (a block section or a path component) the train has just passed.

- "Request infrastructure resource" event: once a train arrives at a request point, a "request infrastructure resource" event will be triggered. The OCC will carry out a conflict-free test and a deadlock-free test. If the train is allowed to occupy the requested infrastructure resource, a "grant movement authority" event will be triggered. Otherwise, the request will be put into a pending list of requests.

- "Release infrastructure resource" event: the infrastructure resource specified in the event will be released as free. The OCC will check the pending list of requests. If one or more requests are pending due to the occupancy of the just-released infrastructure resource, a suitable request will be selected and the requested infrastructure resource will be granted. The conflict-free and deadlock-free tests should be carried out (see Section 5). A "grant movement authority" event will be triggered afterwards.

The occurrence time of an event should be scheduled to indicate when the event is expected to occur. Similar to time-driven simulation, a system clock is initialised for all the involved trains and infrastructure. The system clock is set to the occurrence time of the first of the scheduled events, according to which the system states are updated. After all the events of the current time of the system clock have been processed, the system clock is set to the scheduled time of the immediately subsequent event(s). Thus the behaviour of the system is simulated over discrete time points successively. It is possible that several events take place at one specific time point. These events should have no interdependency and can be processed in an arbitrary sequence. Otherwise, the events should be scheduled in several steps, although these events take place at the same time point. For example, a "request infrastructure resource" event can only be scheduled after the "train arrive request point" event has been processed. In contrast with time-driven simulation, the time-intervals between two successive time points are variable in event-driven simulation. The interdependency of events is shown in Figure 3-3. The dashed line represents the possibility of another event being triggered after a given event has been processed.

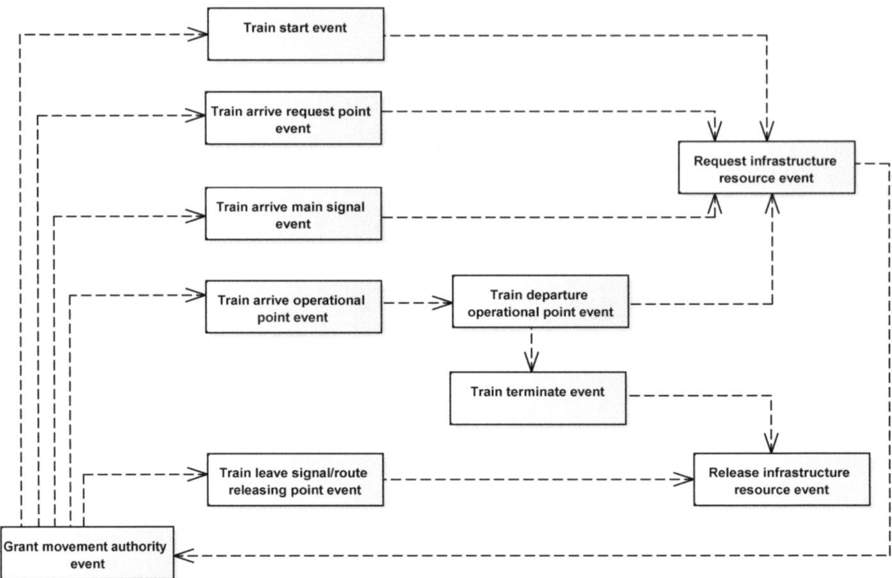

Figure 3-3: Interdependency of events

The workflow of event-driven simulation is shown in Figure 3-4. A event queue is maintained to manage events among the involved trains and infrastructure according to the classical observer pattern (Gamma 1995). The occurrence time of the first-most event is derived from the event queue and is set as the current time of system clock. The related events will be dispatched to the observers, e.g. a train or the OCC, which are responsible for the handling of the events. For example, a "request infra-structure resource" event will be raised by a train and be dispatched to the OCC, and a "grant movement authority" event will be dispatched afterwards to the requesting train. Accordingly, the tasks of requesting and allocating infrastructure resources are distributed to trains and the OCC through the dispatched events.

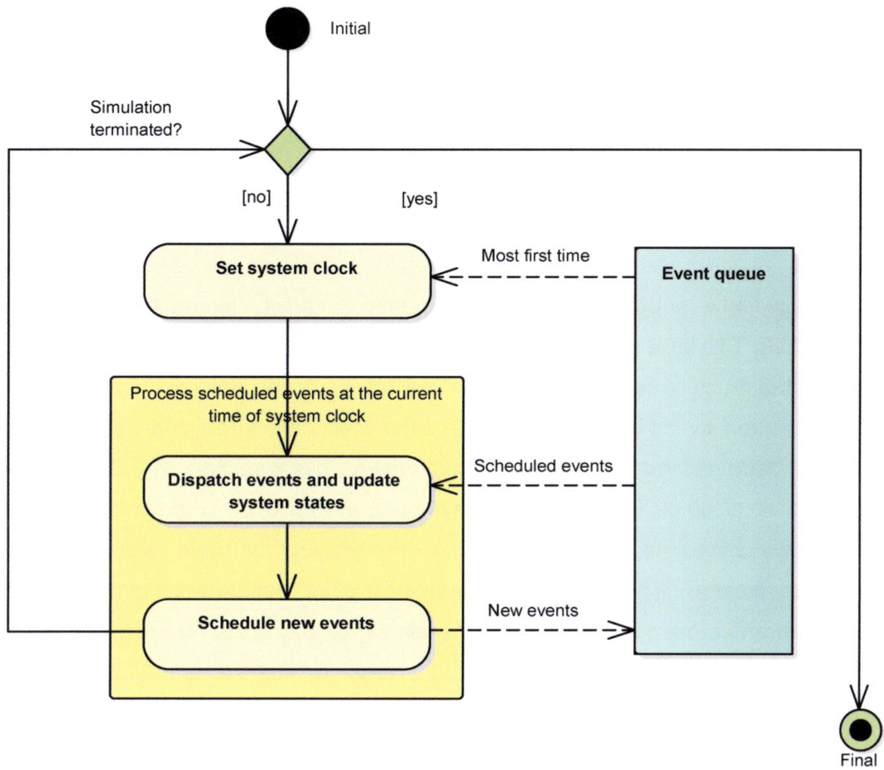

Figure 3-4: Workflow of an event-driven simulation

In time-driven simulation, the position of trains at each fixed time point should be forecasted at the beginning of the single processing step for each time interval, while in event-driven simulation the position and the corresponding time point of a train are scheduled once a "grant movement authority" event is dispatched to the train. According to the position arrived at (e.g. signals or stops) and the time points, new events will be scheduled and inserted into the event queue.

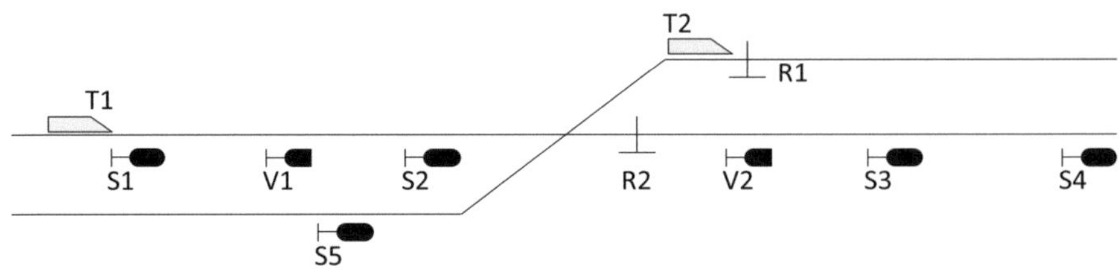

Figure 3-5: An example of event-driven simulation

Here a simplified event-driven simulation is carried out for the example shown in Figure 3-5. Train T1 starts its train run from main signal S1, at which the train raises a "request infrastructure resource" event for the block section from main signal S1 to S2 (represented by S1→S2) at 12:00:00. After the conflict-free and deadlock-free tests have been passed, the movement authority for block section S1→S2 is issued to train T1 at 12:00:00 (To simplify the example, the required communication time, the signal watching time, the reaction time for drivers, and the route setting/releasing time are set as zero). Then, a "train arrive request point" event (for distant signal V1) and a "train arrive main signal" event (for main signal S2) are scheduled by train T1 at times 12:01:00 and 12:02:00. At 12:01:00 a "request infrastructure resource" event (for block section S2→S3) is triggered by train T1. The request is not approved because of the occupancy conflict caused by train T2. The request will be put into the pending list of requests. Train T2 is supposed to leave the route releasing point at 12:02:30. The "train arrive main signal" event for train T1 at main signal S2 is triggered at 12:02:00. At this moment, train T1 has to wait for the issue of movement authority for block section S2→S3. The next scheduled events are the "train leave route releasing point" event and the "release infrastructure resource" event for train T2 at 12:02:30. Train T1 is approved to occupy block section S2→S3 at 12:02:30. It then schedules the events for the new movement authority issued by the OCC. The list of event queue at 12:02:30 is shown in Table 3-1.

Scheduled time	Event	Observer	Infrastructure
12:00:00	Request infrastructure resource	OCC	S1→S2
12:00:00	Grant movement authority	T1	S1→S2
12:01:00	Train arrive request point	T1	V1
12:01:00	Request infrastructure resource	OCC	S2→S3
12:02:00	Train arrive main signal	T1	S2
12:02:30	Train leave route releasing point	T2	R1
12:02:30	Release infrastructure resource	OCC	S5→R1
12:02:30	Grant movement authority	T1	S2→S3
12:03:30	Train leave route releasing point	T1	R2
12:04:00	Train arrive request point	T1	V2
12:05:00	Train arrive main signal	T1	S3

Table 3-1: The list of event queue at 12:02:30

It should be noted that the event queue is developed dynamically during the evolution of the simulation. At 12:02:30, the events for train T1 are scheduled at 12:03:30, 12:04:00, and 12:05:00 respectively. It is possible for new events to be inserted and scheduled events modified according to the operational situation. For example, a "request of infrastructure" event will be inserted at 12:04:00, which is triggered by the "train arrive request point" event (for distant signal V2). If the requested block section S3→S4 can be granted immediately without any conflicts, the "grant movement authority" event will be scheduled for the new block section S3→S4. Since the running time for arriving at main signal S3 is scheduled based on the assumption that the train will be stopped at signal S3, the occurrence time of the scheduled events should be recalculated for the newly granted movement authority. The arrival time at main signal S3 is thus shifted from 12:05:00 to 12:04:30 due to the hindrance-free train movement from distant signal V2 to main signal S3. The list of event queue at 12:04:00 is shown in Table 3-2, in which the modified and the newly inserted events are underlined.

Scheduled time	Event	Observer	Infrastructure
Before 12:02:30	...		
12:02:30	Grant movement authority	T1	S2→S3
12:03:30	Train leave route releasing point	T1	R2
12:03:30	Release infrastructure resource	OCC	S2→R2
12:04:00	Train arrive request point	T1	V2
12:04:00	Request infrastructure resource	OCC	S3→S4
12:04:00	Grant movement authority	T1	S3→S4
12:04:30	Train arrive main signal	T1	S3
12:06:30	Train arrive main signal	T1	S4

Table 3-2: The list of event queue at 12:04:00

The exact interactions between trains and infrastructure can be realistically modelled and simulated through event-driven simulation, which is widely used for railway simulation. The most two important tasks in an event-driven simulation are the calculation of running dynamics and the simulation of the signalling system. With running time calculations, each time point of the events can be scheduled in advance, since the train can run under the granted movement authority without unexpected hindrance. Through interaction with the signalling system, the train can receive a new movement authority, with which a new round of calculation of running dynamics will be initiated.
In Section 4, the workflow of calculating running dynamics and the simulation of signalling systems for an individual train run will be presented based on event-driven simulation.

3.3   Timetable Simulation and Operational Simulation

In railway simulation, the randomness of railway operations, including factors such as failures of infrastructure and rolling stocks, accidents, stochastic behaviour of passengers, as well as other internal and external variations during railway operations, should be taken into consideration. Depending on the modelling approach and the randomness in railway operations, railway simulation can be categorised as timetable simulation as well as operational simulation.

During timetable simulation, railway operations are considered to be deterministic processes, in which the initial values and the parameters are fixed. The applications of timetable simulation are:

- Construction of timetables (with asynchronous simulation)
- Identification of conflicts for a given timetable or a specific operating program
- Calculation of scheduled waiting time and synchronisation time
- Determination of capacity of a given network for a specific operating program
- Identification of bottlenecks

The workflow of a timetable simulation can be either time-driven or event-driven (see Section 3.2). Random disturbances are considered a fixed input. Train dynamics are calculated deterministically in a timetable simulation with fixed parameters. In order to ensure punctuality and stability of railway operations, the recovery time and the buffer time are provided in timetables to overcome the influences of a given disturbance. To achieve a high level of accuracy, the parameters used for timetable simulation should be calibrated, the process for which will be discussed in Section 6.3.

With operational simulation, randomly generated stochastic disturbances are introduced in order to investigate the quality and the robustness of a specific operating program. Multiple rounds of simulations will be carried out; with different randomly generated inputs of disturbances for each round. The statistical attributes will be further analysed based on the results of multiple rounds of simulations. The applications of operational simulation are:

- Determination of the quality of service for a given timetable
- Analysis of delays for specific train groups
- Examination of the stability, robustness, and sensitivity of a timetable against certain types of disturbances or events (e.g. in construction sites)
- Analysis of delay propagation and the relationship between primary delays and consecutive delays
- Investigation of the possibility to recover from delays
- Determination of the deviation between the scheduled and simulated operations

The same workflow and algorithms for the calculation of running dynamics and conflict resolution can be applied for both timetable simulation and operational simulation. Timetable simulation can be regarded as a special case of one round of simulation, in which the disturbances are set as zero. For operational simulation, the random disturbances should be modelled and generated, which leads to different simulation results in each round of simulation.

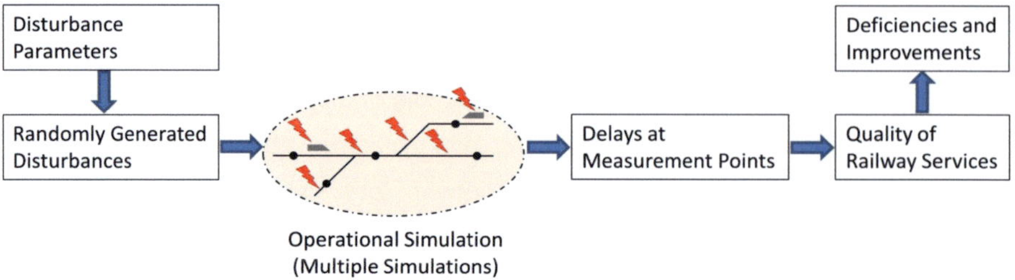

Figure 3-6: Disturbances and delays in operational simulation (Cui et al. 2016)

Possible disturbances and delays in operational simulation are shown in Figure 3-6. At first, the probability distribution of disturbances should be determined, which is used to model the stochastic disturbances. Then the randomly generated disturbances can be generated and introduced into an operational simulation. The delays at each measurement point will be recorded for evaluation of the quality of the railway services. Thereafter, the existing deficiencies and the possible improvements of the investigated railway system can be derived.

Either statistic distributions or empirical distributions can be applied to model the disturbances. If the statistic distribution of a certain type of disturbance is known, the disturbances will be produced from the mathematical equation directly. For example, the disturbances (in the form of time extensions) can be approximated by the negative exponential distribution, the Erlang-K distribution, the lognormal distribution or the Weibull distribution (Yuan 2006). In this case, the disturbances can be generated by the known distributions with certain parameters in the defined distribution function, which are known as **disturbance parameters**. One of the most challenging tasks for operational simulation is the calibration of these disturbance parameters (see Section 6.4). Empirical distributions are modelled based on the available data observed in reality. In case no theoretical distribution can adequately fit the data of the investigated disturbances, the observed data can be used to generate stochastic disturbances directly.

For a determined statistic or empirical probability distribution, random disturbances x (in the form of time extensions) are generated through inverse transformation. With inverse transformation, the cumulative distribution function $C(x)$ of the disturbances x will be derived initially. Then the uniformly-distributed random numbers z ($z \in [0, 1]$) are created. For any random number z_i, the sampling value of x_i can be determined through $x_i = C^-(z_i)$. Since the random disturbances are derived through the inverse function of the cumulative distribution, the method is therefore called "inverse transformation".

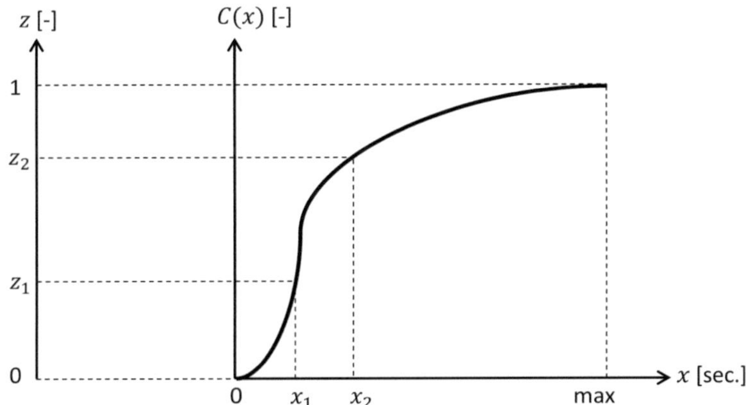

Figure 3-7: Inverse transformation to generate random disturbances

In Figure 3-7, the disturbances x_1 and x_2 are obtained from the random numbers z_1 and z_2. Here the random numbers z ($z \in [0, 1]$) are actually pseudo-random numbers generated in computer systems. A set of random seeds is used to initialise a pseudo-random number generator with a certain algorithm, e.g. Linear Congruential Generators (LCG) developed by (Lehmer 1951). For operational simulation, it is very common to repeat the generation of the same random numbers to analyse the effects with different variants. Therefore, random seeds will be defined in advance to control the randomness during operational simulation.

4 Simulation of an Individual Train Run

4.1 Calculation of Running Dynamics

4.1.1 Numeric Calculation and Workflow of Calculating Running Dynamics

Once a movement authority (see Section 3.2.2) is granted to a train, the running dynamics that describe the exact trajectory of the train for the new granted movement authority should be calculated. The basic information of running dynamics includes the current time, the (head) position of the train, as well as the velocity of the train at the given time point. The tractive force and the resistance of the train are necessary for calculation of running dynamics. Hence the resulting acceleration and the consumption of energy can be determined. Through the calculation of running dynamics, the train movement and the occurrence time of each event can be predicted exactly and scheduled, and the train run can be simulated according to the scheduled events. The calculation of running dynamics is fundamentally based on Newton's laws. For the continuous changing of train movements, it is practical to carry out the calculation with numerical methods, by which the trajectory of train movements can be derived upon in many discrete steps. A series of **discrete point**s can be used to describe the running dynamics of the entire train run. Each discrete point contains the following information:

- Distance from the last discrete point to the current point
- Running time from the last discrete point to the current point
- Velocity at the current point

Among others, three different approaches, time-step, distance-step, and velocity-step, can be applied for the calculation of running dynamics typically. With the time-step approach, running dynamics are calculated at each fixed time interval, making it suitable for a time-driven simulation (see Section 3.2.1). By distance-step approach, the running dynamics can be derived along the path of the train step by step with a fixed distance. The time-step approach and the distance-step approach are not efficient for a situation in which a train is running with constant speed. Moreover, the results of the time-step approach and distance-step approach depend on the interval size. They can change greatly with the variation of velocity. Therefore, the time-step approach and the distance-step approach are not as efficient as the velocity-step approach.

The velocity-step approach for calculation of running dynamics is suitable for event-driven simulation, and conforms naturally to the changes of train movements and running dynamics. The density of discrete points increases with the rate of the speed change and can therefore exactly reflect the change of running dynamics in the acceleration and braking phases. Using the velocity-step method, the efforts of calculation for a train running at a constant velocity are reduced, which can improve the system performance of running time calculation in the cruising phase.

Based on the modelled infrastructure, rolling stocks, and the operational conditions (see Chapter 2), the calculation of running dynamics is carried out for each discrete step (for example, in a velocity-step). In (Brünger, Dahlhaus 2014) and (Neuber 2017), the calculation of the tractive force, resistances, accelerations, as well as the running dynamics are summarised and presented. In this chapter, the workflow of the numerical calculation of running dynamics is the point of focus.

During a simulation process, a round of calculation of running dynamics is carried out for the section from the actual position of the head of the train to the EOA. This section is defined as an **authorised running section**. To reduce the complexity of computation, it is necessary to divide the whole authorised running section into several sections, by which the computation task for an authorised running section can be separated into several independent parts. The definition of these sections is presented in Section 4.1.2. Afterwards, the entry velocity and exit velocity should be set as pre-conditions for further detailed calculation (see Section 4.1.3). The detailed calculation of running dynamics will be carried out finally, providing exact outputs of the position and the velocity at each time point (see Section 4.1.4). The overview of the workflow of calculation of running dynamics is shown in Figure 4-1.

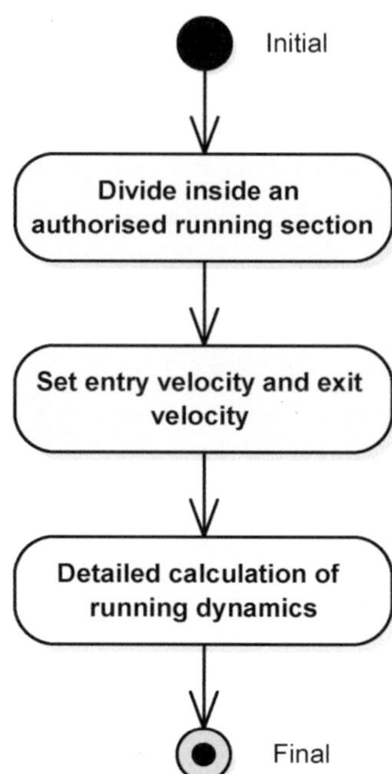

Figure 4-1: Overview of the workflow of the calculation of running dynamics

In this chapter, only the algorithm for calculation of running dynamics with minimum running time is discussed. However, the presented workflow of the calculation of running dynamics can be applied for all driving styles. The calculation of minimum running time is the basis for various driving styles, for example, the driving style with optimised energy consumption (see Appendix B).

4.1.2 Definition of Sections inside an Authorised Running Section

During the simulation process, the running dynamics are calculated inside the given authorised running section. For the calculation of running dynamics, an authorised running section can be further divided into moving sections, characteristics sections and behaviour sections (Brünger, Dahlhaus 2014).

A **moving section** represents an independent part for the calculation of running dynamics inside of an authorised running section. It is limited by the current train position, a stop with a scheduled dwell time, or the EOA. An authorised running section can be covered completely by one or several moving sections without gaps or overlapping. If several moving sections are covered in an authorised running section, there is a stop with a scheduled dwell time between two neighbouring moving sections. Therefore the calculation of running dynamics for each moving section can be carried out independently.

Within a moving section, one or several characteristic sections are defined. A **characteristic section** describes a part of infrastructure with identical geometrical attributes and maximum allowed velocity. Hence the calculation of running dynamics for a characteristic section can be handled with at most one full operational cycle of behaviour sections.

A full operational cycle of behaviour sections includes a sequence of behaviours: acceleration, cruising, coasting, and braking (Brünger, Dahlhaus 2014). The term **behaviour section** is a general term used to describe the behaviour (acceleration, cruising, coasting, and braking) at one characteristic section. At the behaviour section of acceleration, the value of acceleration of the train is positive. At the behaviour section of cruising, the train runs with constant speed, and the value of acceleration is zero. At the behaviour section of coasting, neither tractive force nor braking force is applied, and at the behaviour section of braking, braking force is applied. The value of acceleration of the train for the braking section is negative.

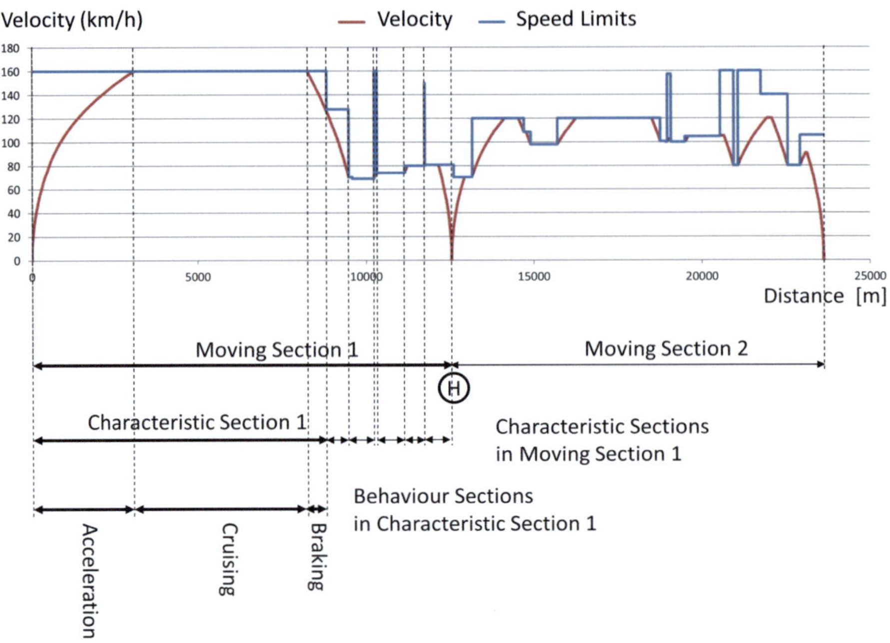

Figure 4-2: An example of moving sections, characteristic sections and behaviour sections

An example of moving sections, characteristic sections and behaviour sections is shown in Figure 4-2. To illustrate the principles of the dividing sections, an extremely long authorised running section with a length of 23.6 km is considered in the example. Inside the authorised running section, two moving sections (moving section 1 and moving section 2) are included. There is a stop (marked with the sign "H") with a scheduled dwell time between the two moving sections. In Figure 4-2, the characteristic sections in moving section 1 are marked. Each characteristic section has an identical maximum allowed velocity and geometrical attributes. For example, the maximum allowed velocity of characteristic section 1 is 160 km/h. The characteristic section 1 can be further divided into several behaviour sections. It is not necessary that a characteristic section include all the possible behaviour sections. The resulting behaviour sections depend on the applied driving style. In this example, the train is running with minimum running time. Therefore, only three behaviour sections, acceleration, cruising, and braking, are included in characteristic section 1.

After calculating running dynamics, the exact trajectory will be saved in each behaviour section in the form of a series of discrete points (see Section 4.1.1). The class diagram of sections and discrete points is shown in Figure 4-3.

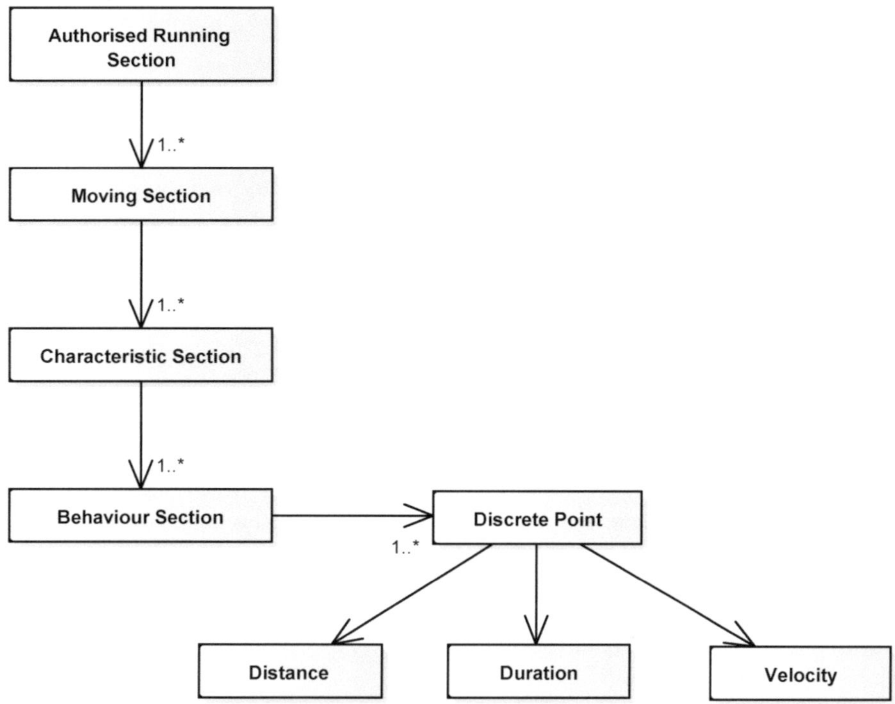

Figure 4-3: Class diagram of sections and discrete points

4.1.3   Setting Entry Velocity and Exit Velocity

Before the running dynamics for a moving section are calculated in detail, the entry velocity and exit velocity of the characteristic sections within the moving section should be determined, in order to ensure the train can stop before the EOA or a stop without exceeding the maximum allowed velocity. Therefore, a backwards setting of entry/exit velocity (Brünger, Dahlhaus 2014) is carried out with consideration of braking behaviour at each characteristic section.

The characteristic sections, which could have braking behaviour, should be initially identified as "with-braking sections". That is to say, if the maximum allowed velocity of a characteristic section is larger than that of the next characteristic section; it is categorised as a with-braking section. The last characteristic section being considered is always a with-braking section, since the train is supposed to stop at the end of the moving section.

Starting from the last characteristic section, the exit velocity V_{exit} of the characteristic section is set as zero (as the train is supposed to stop). The maximum allowed entry velocity V_{entry} will be calculated with the consideration of the maximum allowed velocity V_{max} and the length of the characteristic section. If the characteristic section is long enough for the train to brake from V_{max} to V_{exit}, the entry velocity is set as V_{max}. Otherwise, it will be reduced until the required distance for braking from V_{entry} to V_{exit} is less than the length of the characteristic section. The calculated entry velocity V_{entry} will also be set as the exit velocity of the previous characteristic section.

The process will then be repeated backwards. If the current characteristic section is not a with-braking section and the exit velocity has not been set yet, the exit velocity will be set as V_{max}, since there is no braking behaviour required. If the current characteristic section is a with-braking section, the exit velocity of the current characteristic section will be set as either the entry velocity of the next characteristic section (if it has already been calculated in advance) or the maximum allowed velocity V_{max}. The entry velocity should be set to satisfy the constraint that the braking length should be less than the length of the characteristic section.

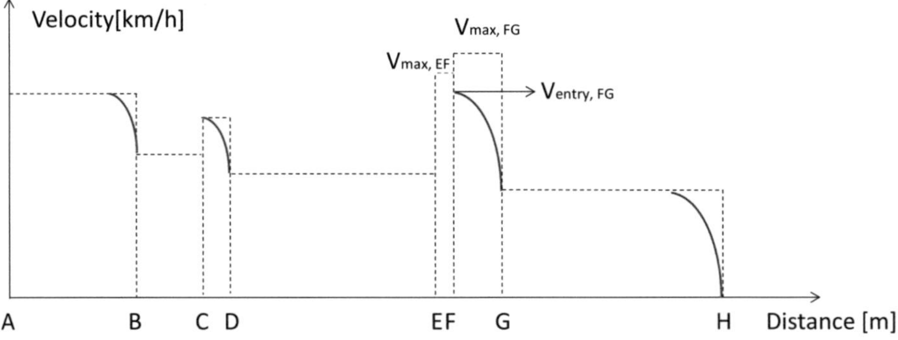

Figure 4-4: An example of backwards setting of entry velocity and exit velocity

In Figure 4-4, the entry velocity and exit velocity are set backwards. The characteristic sections A-B, C-D, and G-H are with-braking sections, at which the distance for braking from the maximum allowed velocity to the exit velocity is shorter than the length of the characteristic section. The entry velocity is set as the maximum allowed velocity. The characteristic section F-G (a with-braking section),is not long enough for braking from the maximum allowed velocity $V_{max,FG}$ to the exit velocity and the entry velocity should therefore be reduced to $V_{entry,FG}$. At this step, the exit velocity of the characteristic section E-F is also set as $V_{entry,FG}$.

During the backwards calculation, the braking behaviour is considered in order to ensure that the train can stop at the end of the moving section. The maximum allowed entry velocities and exit velocities of the characteristic sections are calculated. However, it is possible that the train cannot reach the exit velocity due to low entry velocity or the short length of the characteristic section. Therefore, a forwards setting of entry/exit velocity is required.

In contrast with backwards setting, the process of forwards setting starts at the first characteristic section. The initial velocity of the moving section is set as the entry velocity of the first characteristic section. If the length of the current characteristic section is insufficient, the train may not able to reach the given exit velocity. The exit velocity of the characteristic section and the entry velocity of the next characteristic section should be then reduced accordingly. Otherwise, the value of the entry velocity and the exit velocity that were calculated in backwards calculation can be kept.

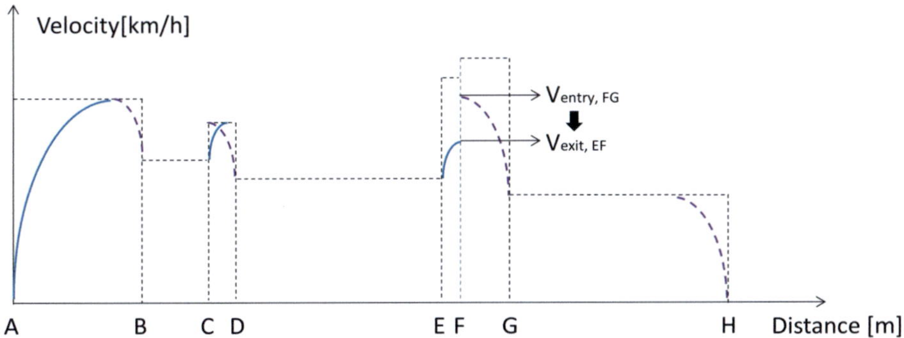

Figure 4-5: An example of forwards setting of entry velocity and exit velocity

The forwards setting of the entry velocity and the exit velocity is shown in Figure 4-5. For the characteristic section E-F, it is not possible to reach the entry velocity $V_{entry,FG}$ with full acceleration. Therefore the exit velocity of characteristic section E-F and the entry velocity of the characteristic section F-G should be set as $V_{exit,EF}$. For the characteristic sections A-B, C-D, and G-H, the entry/exit velocity that were set in backwards calculation are not changed.

4.1.4   Detailed Calculation of Running Dynamics

After setting the entry velocity and exit velocity, the behaviour sections of each characteristic section can be established in the form of a series of discrete points.

To establish behaviour sections in a characteristic section, the possible reached maximum velocity V_{reach} should be determined firstly. With the velocity-step approach (see Section 4.1.1), the check of the possible reached maximum velocity will be started from the maximum allowed velocity V_{max} of the characteristic section. The acceleration curve (from V_{entry} to V_{max}) and the braking curve (from V_{max} to V_{exit}), as well as the required length for acceleration ($s_{acceleration}$) and for braking ($s_{braking}$) are calculated initially. If the length of characteristic section is longer than the sum of $s_{acceleration}$ and $s_{braking}$, the possible reached maximum velocity V_{reach} can be set as V_{max}. Otherwise, the possible reached maximum velocity should be reduced with the interval of the velocity $V_{interval}$ used in the velocity-step approach. This procedure will be repeated until the possible reached maximum velocity is attained, with which the length of the characteristic section is long enough for acceleration and braking.

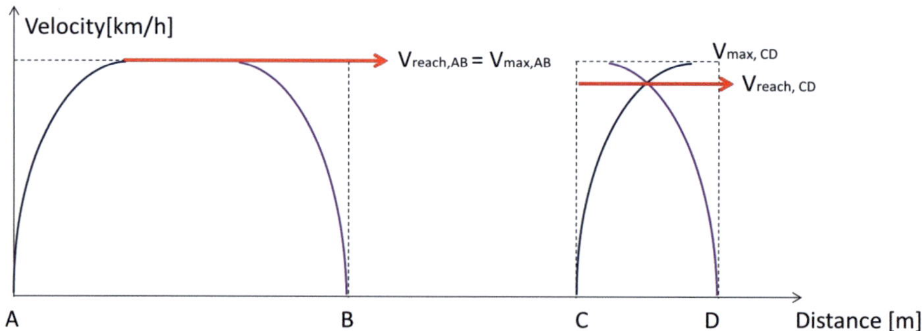

Figure 4-6: Setting of maximum possible reached velocity

In Figure 4-6, the setting of the possible reached maximum velocity for two characteristic sections is illustrated. If the length of the characteristic section A-B is long enough for acceleration and braking, the possible reached maximum velocity is set as $V_{max,AB}$. For the characteristic section C-D, the possible reached maximum velocity $V_{reach,CD}$ should be reduced, since the length of the characteristic section C-D is too short to accelerate and brake at the maximum velocity $V_{max,AB}$.

Once the possible reached maximum velocity V_{reach} is determined, the acceleration curve can be constructed. For calculating running dynamics with minimum running time, the braking curve, as well as the behaviour section for cruising, can also be determined. The length of the behaviour section for cruising ($s_{cruising}$) is the difference between the length of the characteristic section and the sum of $s_{acceleration}$ and $s_{braking}$.

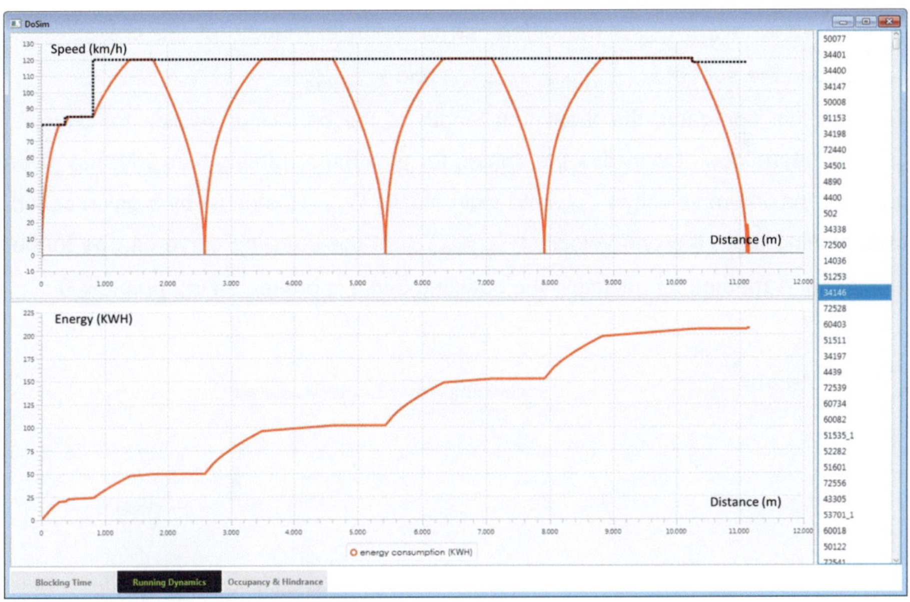

Figure 4-7: An example of the calculated speed profile and energy consumption with minimum running time in software PULSim

The results of the calculated running dynamics will be recorded in the form of a series of discrete points, with which the running dynamics and the energy consumption can be calculated. An example of the calculated speed profile and the energy consumption for a train run with minimum running time in software PULSim is shown in Figure 4-7.

For other driving styles (for example, a driving style with optimised energy consumption), the behaviour section for coasting should be calculated. The initial velocity of the behaviour section for coasting is the possible reached maximum velocity V_{reach}. At first, the velocity at the end of the behaviour section for coasting is set as the exit velocity V_{reach} of the characteristic section. The length of the behaviour section for coasting ($s_{coasting}$) can be calculated. If the sum of $s_{acceleration}$, $s_{braking}$, and $s_{coasting}$ is less than the length of the characteristic section, the whole behaviour section can be determined. Otherwise, the velocity at the end of the behaviour section for coasting will be increased with $V_{interval}$, until the sum of the $s_{acceleration}$, $s_{braking}$, and $s_{coasting}$ is less than the length of the characteristic section. Here the value of $s_{braking}$ is calculated from the velocity at the end of the behaviour section for coasting to V_{exit}, and the value of $s_{cruising}$ is the difference between the length of the characteristic section and the sum of $s_{acceleration}$, $s_{braking}$, and $s_{coasting}$.

Through this approach, the maximum length of the behaviour section for coasting can be determined. It provides the values for maximum running time with the given possible maximum velocity V_{reach} and exit velocity V_{exit}. Restricted by a given scheduled running time, the end velocity $V_{coasting,end}$ of the coasting curve can be further increased. A method for adjusting the coasting curve is presented in Appendix B.

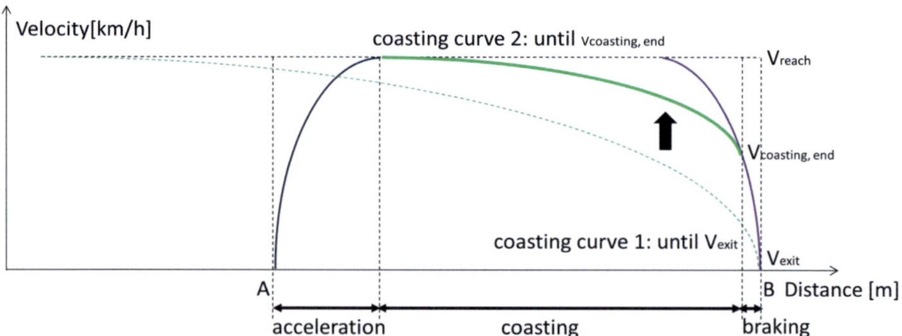

Figure 4-8: Setting of the behaviour section for coasting

An example of setting a coasting curve is shown in Figure 4-8. In the case of coasting curve 1, which ends at the velocity of V_{exit}, the length of the coasting curve is longer than the length of the characteristic section A-B, making it impossible to be implemented. The end velocity of the coasting curve is increased until $V_{coasting,end}$, at which the sum of the length of the behaviour sections for acceleration, coasting, and braking is less than the length of the characteristic section A-B. Therefore, coasting curve 2 will be used for building behaviour sections.

With optimised energy consumption (see Appendix B), the calculated speed profile is shown in Figure 4-9. Compared to the results shown in Figure 4-7, the coasting phase is implemented with the driving style for optimised energy consumption. Accordingly, for the same train run, the total energy consumption is reduced from 207 KWH (for the driving style with minimum running time) to 148 KWH.

Figure 4-9: An example of calculated speed profile with optimised energy consumption in software PULSim

4.2 Simulation of Signalling Systems

4.2.1 Functions of a Signalling System

During the simulation of a signalling system, the safety functions used in railway operations to achieve an appropriate train separation are simulated. The safety functions include both infrastructure functions and train-borne functions.

Through simulating infrastructure functions, the status of the signalling system (e.g. the current position of turn outs, the aspect of signals) is collected, and the signalling system is controlled through the OCC and interlocking systems. The requests of occupying infrastructure resources are managed in time-driven or event-driven simulations (see Section 3.2). A train can only be allowed to occupy the requested infrastructure resources with a granted permission in the form of a movement authority. Conflict situations should be identified and resolved to ensure conflict-free and deadlock-free operations (see Chapter 5). Once an occupied infrastructure resource is released, the status of the infrastructure resource will be updated in time. In addition, the wayside communication systems and interlocking systems will also be managed through infrastructure functions.

Train-borne functions simulate train movements inside the authorised running section, depending on a certain signalling system equipped on the train itself. Although there are many different implemented signalling systems in railway operations, trains can be separated either by fixed block distance or moving blocks in general. In Sections 4.2.2 and 4.2.3, the workflow of simulating train movements with different approaches for train separation is explained in detail.

Each train should calculate its own speed profile to ensure it can be stop or reduce velocity in the rear of the danger point. The danger point is defined as "the location beyond the EOA that can be reached by the front of the train without creating a hazardous situation" (Giaccone, Pomposiello 2011). A dynamic speed profile will be calculated with considerations of detailed running time and braking curves (see Section 4.1). An example of updating a speed profile is shown in Figure 4-10. Before the train reaches point V (a distant signal), it maintains a speed profile for the service brake before signal S1. Once the train has received a new movement authority, the speed profile is updated for the service brake before signal S2. In the case of an event-driven simulation, the events that take place in the train run between S1 and S2 should be also scheduled at this moment.

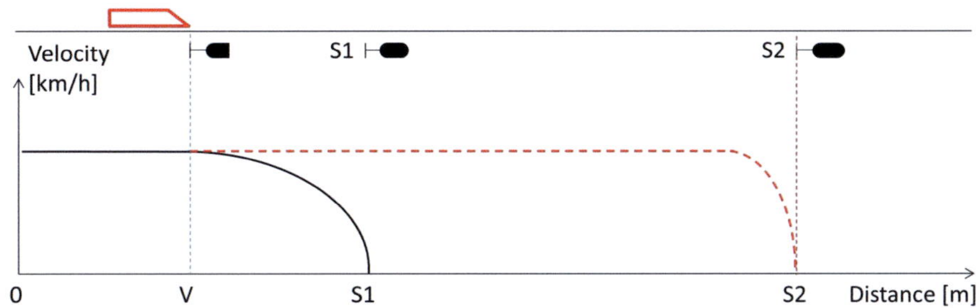

Figure 4-10: Update of the speed profile

It is usually assumed that a train will follow the calculated speed profile in a simulation tool. Except for the applications for the simulation of the stochastic behaviour of train drivers, the functions of supervising train speed as well as the warning/intervention of train running are not necessarily simulated in a simulation tool, although they are the core functions of Automatic Train Protection (ATP) systems.

4.2.2 Simulation of Train Run with Fixed Block Distance

The interaction between the infrastructure side and train side of a signalling system will be simulated depending on the applied ATP system. For fixed block distance, ATP systems can be basically categorised as intermittent and continuous ATP systems, according to their method of data transmission. With intermittent ATP systems (e.g. PZB in Germany, Automatic Warning System (AWS)/Train Protection and Warning System (TPWS) in United Kingdom, ETCS in level 1), the information related to target speeds and the aspects of the approached signals are provided to the train at several discrete points only, e.g. from the mounted balises between the rails. With a continuous ATP system (e.g. LZB in Germany, ETCS in lever 2), movement authorities are updated and sent continuously to trains via the communication system.

A request point is used to schedule a request for the next block section. Once the position of a request point has been determined, the timing of requesting the next block section can be scheduled, with the consideration of signal watching time and route setting time. In order to demonstrate the principles for requesting a block section, the signal watching time and the route setting time are not included in the following simplified examples in Section 4.2.2.

For an intermittent ATP system, the timing for requesting the next infrastructure resource (block section) should be scheduled before arriving at the related distant signal or the combination signal. For example, a "train arrive request point" event can be scheduled for the current movement authority, in order to apply to the next block section related to the distant/combination signal inside the authorised running section. An example of the simulation of a train run with an intermittent ATP system is shown in Figure 4-11. The movement authority for the block section S1→S2 is granted at point A, which is the combination signal S1. The block section will be released at point B, if the rear of the train passed the end of the overlap behind signal S2.

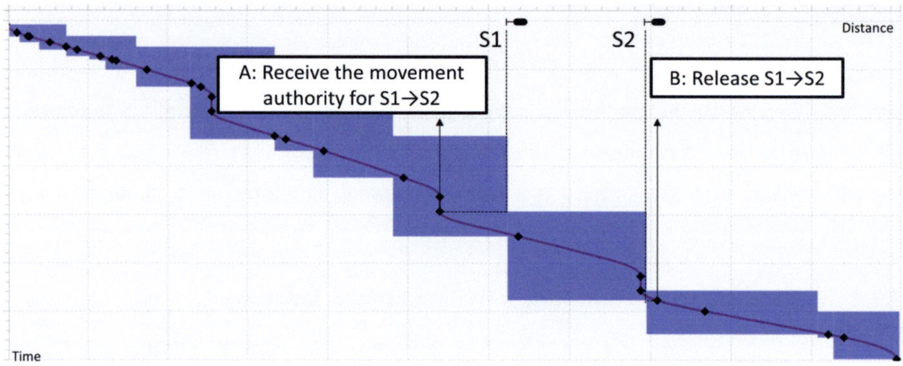

Figure 4-11: Simulation of a train run with intermittent ATP including blocking time stairway

Depending on the velocity of the train, the request point for the next block section with intermittent ATP is related to a fixed point (e.g. before a distant signal or a combination signal). On the contrary, in a continuous ATP system, the request point is determined dynamically according to the speed profile. A fixed request point is not available in a continuous ATP system. The request point should be located at the point at which the train should be able to safely brake to a stop if the next block section has not been granted yet, or to reduce velocity to a certain speed limitation if another train run is in front.

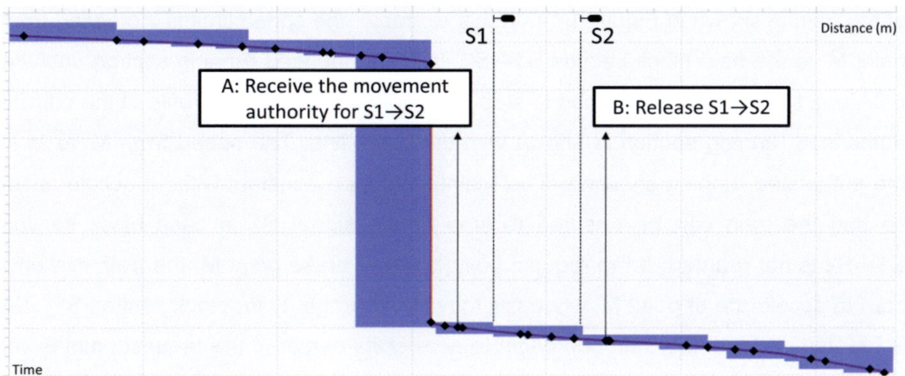

Figure 4-12: Simulation of a train run with continuous ATP including blocking time stairway

In Figure 4-12, an example of the simulation of a train run with continuous ATP is shown. At point A, the movement authority for block section S1→S2 is granted. Point A is not a fixed point like in an intermittent ATP system (see Figure 4-11). The point for requesting and granting movement authorities should be derived through the calculated trajectory of train runs.

Two different methods can be used to locate the request point. In Figure 4-13, the request point is located at the point M, where the train can stop before signal S1. With this method, the next block section will be granted at the braking point of the authorised running section to avoid unnecessary braking.

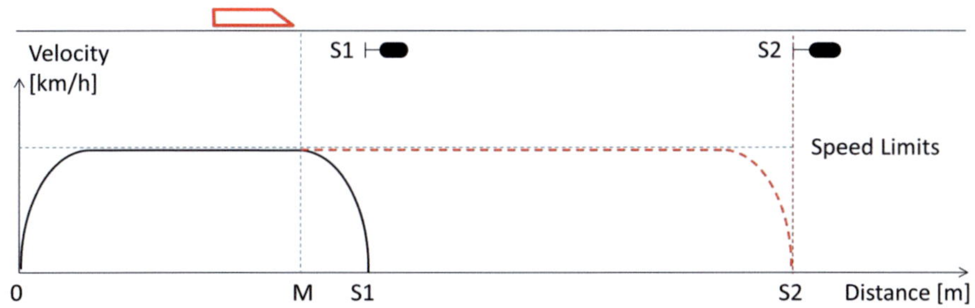

Figure 4-13: Set a request point at the braking point for continuous ATP

However, this method of locating the request point at the braking point cannot always guarantee an efficient train run. An example with various speed limits along the path of the train is shown in Figure 4-14. In this example, the speed limit is increased from point M' to the next block section S1→S2. In the authorised running section until the EOA of S1, the train is supposed to stop at signal S1. The speed profile at the current authorised running section is shown with the black line. The speed from M' to M in the authorised running section will be maintained as a constant without acceleration, so that the train can be ensured to stop before signal S1 in case block section S1→S2 is not granted. If the request point is set at brake point M, the train can only start to accelerate at point M, when the movement authority for block section S1→S2 is granted. Indeed, the train can begin to accelerate earlier, if the request point is set at point M'. Therefore, it is more efficient to set the request point at point M', where the current speed profile diverges from the predicted speed profile based on the assumption that the next movement authority is supposed to be granted.

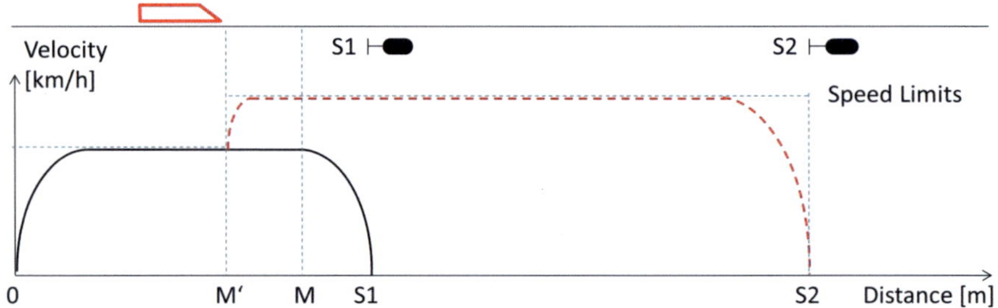

Figure 4-14: Set request point at the diverging point for continuous ATP

For both intermittent and continuous ATP systems, special attention should be paid to the stops with scheduled dwell times inside an authorised running section. If such a stop is available, the position of the request point should be compared to that of the stop. If the stop is closer to the current EOA, the request point should be set at the stop, since the next block section will only be granted before the train departs from the stop.

The timing for releasing an infrastructure resource can be determined based on the speed profile. The route releasing point and the signal releasing point are firstly searched for within the authorised running section. The position of the train head is examined afterwards, when the train is leaving a route/signal releasing point. If the position of the train head is still inside the authorised running section, the time of releasing the related infrastructure resource (a path component or a block section) can be determined, with considerations of the route clearing time and the releasing time. The same process can be carried out for both intermittent and continuous ATP systems.

4.2.3   Simulation of a Train Run with Moving Block

In modern signalling systems, a more flexible approach of train separation with the use of moving blocks is emerging for the new generation of ATP systems. In rail-based urban public transport, Communication-Based Train Control (CBTC) has already been defined in the IEEE 1474 standard (IEEE 2004) and has been implemented in different countries. For railway main lines, the ETCS in level 3 with moving blocks is still under development. Both CBTC and ETCS in level 3 are continuous ATP systems. The implementation of train detection, communication, and supervision may vary in the different solutions. To simulate train runs with a moving block system, the different implementations will be modelled with the corresponding parameters and settings. In this section, the workflow of the simulation of train runs with moving blocks is presented in principle, without considering the different parameters and settings.

For fixed block systems, the safety distance between two trains is maintained by a fixed signal at the boundary of a block section. In a moving block system, the fixed boundary no longer exists. Through continuously transferring the current positional data from the train (especially the position of the rear end) to the infrastructure, the danger points can be determined in order to ensure another train will never pass them.

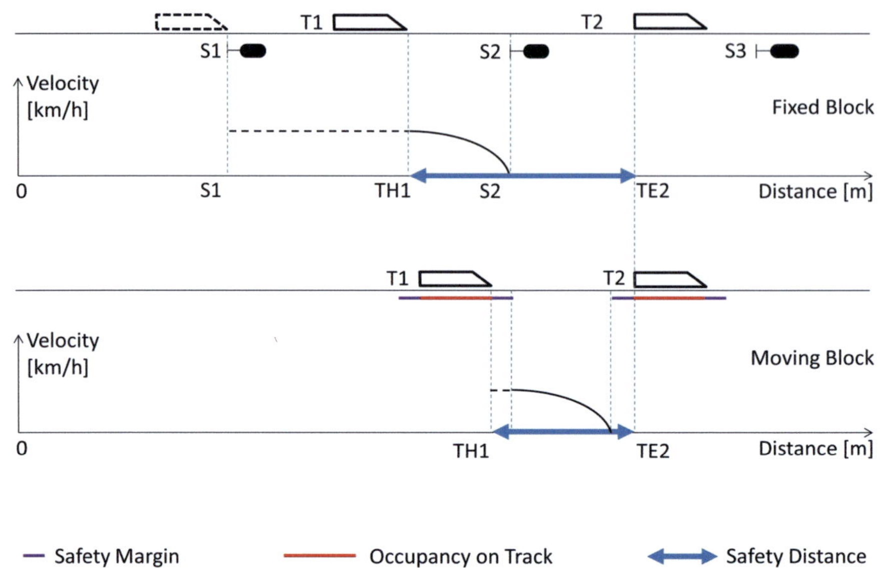

Figure 4-15: Comparison of the safety distance between fixed block and moving block

A comparison of fixed block and moving block systems is shown in Figure 4-15. For the fixed block system, the braking curve for train T1 was calculated at the position of S1. The authorised running section for T1 ends at the signal S2. Therefore, the braking distance starts from TH1 (the head of the train T1) and ends at signal S2. At this moment, the safety distance between train T1 and train T2 is the sum of the braking distance and the distance from S2 to TE2 (the end of the train T2). In a moving block system, the braking curve is dynamically calculated on the train-side through the updated information regarding the danger points. According to the position of train T2, the absolute braking distance from T1 to T2 is calculated. The above-mentioned safety distance is the sum of the braking distance and a certain length of safety margins between the head of train T1 and the rear of the train T2. Normally, the safety distance between T1 and T2 (marked as the distance from point TH1 to TE2) in a moving block system is shorter than that in the fixed blocking system. An optimisation in its value can, therefore, lead to a higher capacity with the implementation of moving blocks.

Although continuous communication and update takes place within a moving block system, it is efficient to schedule discrete timing to request and allocate infrastructure resources during a simulation process. The request point for the next infrastructure resource can be set at the braking point of the authorised running section with the consideration of safety margins. Once the train receives the movement authority of the requested infrastructure resource, further events can be scheduled for the new movement authority. If other trains hinder the concerned train, the information related to the hindrances will be recorded. Once the status of the hindering trains is changed, the request of the hindered train will be examined again, until the train receives a new movement authority.

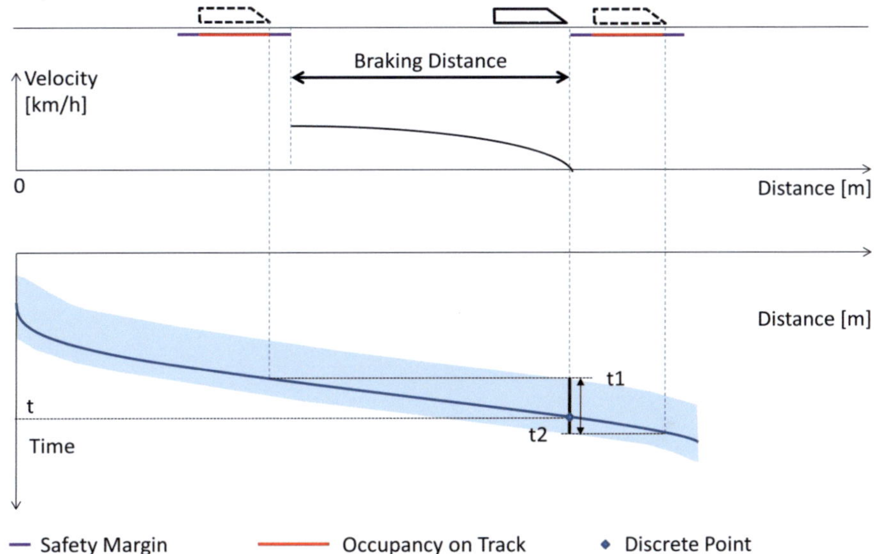

Figure 4-16: Calculation of occupancy time (blocking time) for a discrete point

In contrast with a fixed block system, the shape of the blocking time of a train run with moving blocks does not resemble that of a blocking time stairway. Instead, it forms a continuous time channel. As the precondition for the determination of the blocking time, the occupancy time at a discrete point should be calculated. In Figure 4-16, the approach for calculating the occupancy time is illustrated. At the discrete time point t, the occupancy starts at time point t1, at which the train can be braked to a stop at the current position with a certain safety margin. At time point t2, the rear of the train leaves the current position with a certain safety margin. Hence the occupancy time at time point t can be calculated as the time duration from time point t1 to time point t2. The blocking time of the train run is denoted within the area with a light blue colour. As of the most current version of PULSim, signalling systems with moving blocks have not yet been implemented. The process of determination of blocking time for moving blocks will be presented in Section 7.2.1.

5 Simulation of Multiple Train Runs

5.1 Conflict Resolution

5.1.1 Types of Conflicts in Railway Operations

For a given timetable, multiple train runs will be simulated together, which may produce conflicts. In order to solve any arising conflicts, the train runs should be coordinated within the simulation tools. Even for a scheduled, conflict-free timetable, conflicts can result from deviations of scheduled train runs. The sources of the deviations can be stochastic influences during railway operations or other conflicts. There are six kinds of conflicts (Martin 1995) that may occur during railway operations:

- Occupancy conflict at track sections or block sections
- Occupancy conflict at scheduled stops
- Connection conflicts
- Timetable conflicts
- Dispatching conflicts
- Deadlock conflicts

In railway operations, the usage of infrastructure resources is regulated through the implementation of a signalling system in order to ensure safety. If two or more trains are due to enter the same infrastructure resource, an occupancy conflict will take place at the respective track section, block section, or scheduled stop. Occupancy conflicts are a direct result of stochastic influences, and can lead to the generation of further conflicts. Therefore, occupancy conflicts should be identified (see Section 5.1.2) and resolved (see Section 5.1.3) promptly in a simulation tool.

Connection conflicts arise from deviations of scheduled train runs, due to which the connection of train runs cannot be maintained. Additional dispatching measures are required to ensure the connections and to synchronise the connected train runs.

In principle, a timetable should be conflict-free, however, this is not always guaranteed in railway operations. Timetable conflicts indicate the potential probability of the occurrence of occupancy conflicts caused by deviations of the scheduled timetable due to disturbances (see Section 2.4.5). The timetable conflicts are predicted based on the current operations and the scheduled train runs. With the modelled disturbances and the resulted deviations, any potential timetable conflicts can be predicted and identified.

Depending on the structure and the workflow of dispatching systems, the dispatching decisions issued from different dispatchers and/or dispatching systems may be inconsistent, which may produce dispatching conflicts. Dispatching conflicts can also take place in a simulation tool, if the decisions made by the dispatching system are inconsistent. Furthermore, in multi-scale simulation, dispatching conflicts may occur due to the inconsistencies between microscopic, mesoscopic and macroscopic levels (Liang 2017). In Section 5.1.3, the process of avoid dispatching conflicts in a simulation tool is discussed.

A deadlock situation is a special type of conflict, which may frequently arise in a synchronous simulation. In Section 5.2, the algorithm for deadlock avoidance is presented.

Today, conflicts are still solved manually by dispatchers during railway operations. For a simulation tool, the function of automatic conflict resolution is mainly implemented in the dispatching system. Usually, occupancy conflicts and the timetable conflicts should always be resolved automatically by a simulation tool. Some simulation tools are also capable of solving connection conflicts based on certain rules (e.g. in RailSys, see (RMCON 2010)). If a dispatching system is applied, the simulation tool should also be able to solve dispatching conflicts to avoid inconsistencies of dispatching decisions. For a synchronous simulation tool, deadlocks can be solved manually (e.g. in RailSys) or automatically (e.g. in PULSim).

5.1.2 Identification of Conflicts

Through synchronous simulation, occupancy conflicts should be determined initially. The possible connection conflicts and timetable conflicts can then be identified afterwards. The occupancy conflicts have to be solved punctually, whereas connection conflicts are not absolutely required to be solved. The timetable conflicts will then be predicted according to the modelled stochastic disturbances. Once a dispatching decision is made, the dispatching conflicts are examined (see Section 5.1.3). The algorithm to identify deadlocks will be applied at the time at which infrastructure resources are granted (see Section 5.2).

To determine occupancy conflicts, the blocking time stairways of all train runs will be examined. If there is an overlapping between the blocking time stairways of two train runs, a conflict situation is identified. The nearest occupancy conflict between two train runs, which is known as an **immediate conflict**, is the highest priority concern for a dispatching system. The immediate conflict is defined as an occupancy conflict that takes place at a position at which there is no other occupancy conflict for both conflicting trains.

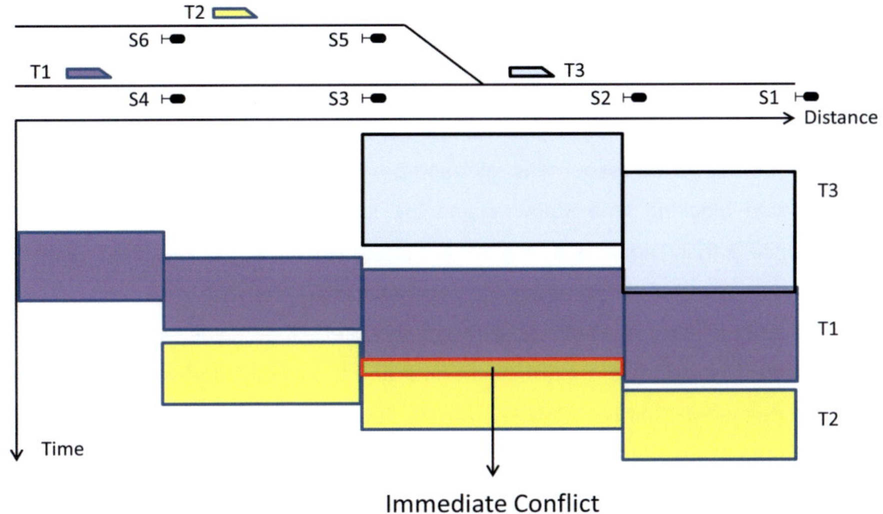

Figure 5-1: An example of immediate conflict

Immediate conflicts are determined through examining the block sections and path components along the path of train. An example of immediate conflicts is shown in Figure 5-1. Along a certain train path, trains T1 and T3 encounter a conflict at the block section S2→S1. However, it is not an immediate conflict. Before the block section S2→S1, another conflict between train T1 and T2 at the block section S3→S2 takes place. Therefore, only the conflict between trains T1 and T2 is considered to be an immediate conflict, and the dispatching system should resolve the immediate conflict punctually. Other conflict situations for non-immediate conflicts may be changed after the dispatching decision is made. For example, if train T1 waits for T2 before signal S3, the conflict between T1 and T3 at the block section S2→S1 no longer exists. Hence it is not necessary to resolve the other non-immediate conflicts at this moment.

The identified immediate conflicts should be recorded and resolved by the dispatching system in time (see Section 5.1.3). The blocking time stairway of each train run should be updated in real-time to identify any potential occupancy conflicts. A method of predicting occupancy conflicts is to use the scheduled speed profile to build a blocking time stairway. However, the actual blocking time of a train run can differ from its scheduled blocking time due to disturbances and the changed running behaviours. Therefore, the predicted blocking time stairway of each train run should be continuously updated to predict the exact conflicts, so that the occupancy conflicts and timetable conflicts can be predicted in time. The times at which the blocking time stairway is to be update are when:

- A movement authority is granted to a train
- An infrastructure resource is released by a train

The updated blocking time stairway and the identified immediate conflicts will be saved. Once a train requests a new infrastructure resource, the identified conflict situation should be resolved immediately (see Section 5.1.3).

Connection conflicts are identified through the predefined rules and thresholds for connections. For example, the transition time and the validity period for train connections are defined in RailSys (RMCON 2010). The transition time defines the required length of time of the feeder train (incoming train) for passengers to transfer. The earliest possible departure time for the connecting train can be derived by adding the transition time to the arrival time of the feeder train. The validity period describes the maximum waiting time of the connecting train. The latest allowed departure time of the connecting train can be also determined by adding the validity period to the scheduled departure time of the connecting train. In case of deviations, the connection between the feeder train and the connection train may be violated, if the earliest possible departure time is later than the latest allowed departure time of the connecting train (see Figure 5-2). By comparing the earliest possible departure time and the latest allowed departure time of the connecting train, the connection conflicts can be identified.

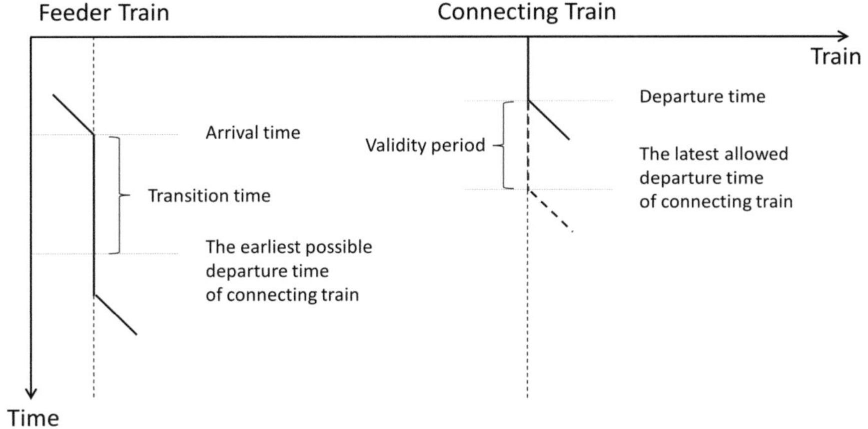

Figure 5-2: Connection conflict between a feeder train and a connecting train

5.1.3 Resolve Conflicts during Railway Simulation

If a conflict situation is identified, a dispatching decision to resolve the conflict should be made by the dispatching systems within a simulation tool. The possible dispatching decisions between two conflicting trains are:

- Apply First-Come First-Served (FCFS) principle
- Apply First-Come Last-Served (FCLS) principle
- Use alternative routes
- Shorten the path of the train
- Cancel the train run

The simplest method of train dispatching is to apply the FCFS principle. With the FCFS principle, the priority of conflicting trains is determined by the requesting time of the requested infrastructure resources. The earlier a train requests an infrastructure resource, the higher the priority that train has. Although the FCFS principle is not always the most efficient way for train dispatching, it is usually implemented in a simulation tool as a "default" dispatching measure. The FCFS principle can at least provide a basis to compare and to evaluate various other dispatching approaches.

If the FCLS principle is applied, the train with a later requesting time can receive a high train priority. This will change the scheduled train sequences. The FCFS and FCLS principles are the most used dispatching measures in an automatic dispatching system. The scheduled path of the train is maintained with both FCFS and FCLS. In the case of both the FCFS and FCLS principle, time-related dispatching (see Section 2.4.5) is applied implicitly, by which the involved train runs may be shifted, and/or the speed profiles may be changed.

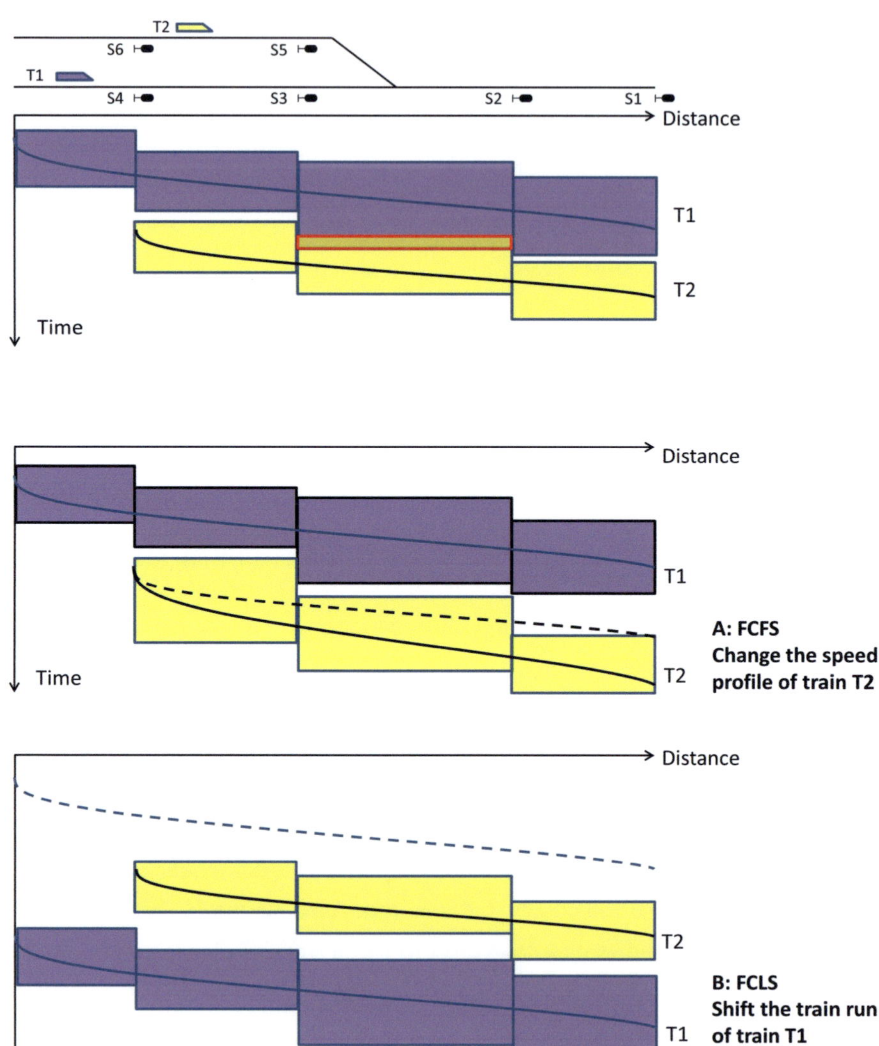

Figure 5-3: Examples of time-related dispatching with FCFS and FCLS

Examples of time-related dispatching are shown in Figure 5-3. Train T1 and train T2 have an occupancy conflict. For a dispatching decision A with FCFS, the speed profile of train T2 is changed so that the conflict can be solved. It is also possible to apply the FCLS principle through shifting the train run of train T1. For time-related dispatching, the resulting deviations may be reconciled via scheduled recovery time, as is the case with conflicts with a small overlapping time. Therefore, experienced dispatchers may ignore these conflicting situations. The function of intelligent prediction of conflicts with consideration of recovery time and regulation of train runs is desirable for an (partially) automatic dispatching system.

If alternative routes for one or both of the conflicting trains are available, the path of the conflicting trains can be changed to apply alternative routes in an automatic train dispatching system. Usually the approach of FCFS, FCLS and the approach with alternative routes can be mixed within a dispatching algorithm.

In case a severe hindrance takes place in the further path of a train, its path can be shortened. In some extreme situations, the train run can even be cancelled. The measures of shortening or cancelling train runs have not yet been widely implemented with current automatic dispatching systems. Therefore, dispatching measures with FCFS, FCLS, and alternative routes are most relevant with automatic dispatching.

There exist five types of possible occupancy conflict situations: following, exiting, merging, crossing, and opposing (Figure 5-4). To solve occupancy conflicts for following or exiting situations, only the speed of the concerned trains must be changed. For trains with merging, crossing or opposing conflicts, a dispatching decision should be made in order to determine train sequences and/or path of trains. For example, in (Martin 1995), decision variables are required for the resolution of conflict situations with merging, crossing or opposing.

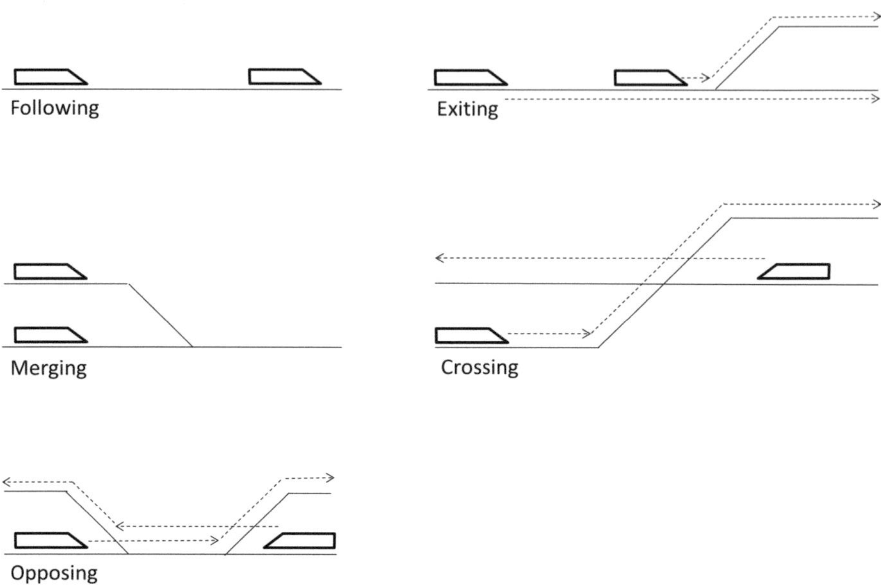

Figure 5-4: Five conflict situations for occupancy conflicts

The final dispatching decision for a conflict depends on the applied dispatching algorithm. For each dispatching decision, the conflicts induced by the decision should also be analysed and evaluated. Usually, a dispatching decision will bring negative impacts to some trains due to the resulting waiting time and/or the changed path of the train. Therefore, the negative impact, which is called malus (Cui et al. 2012), will be compared among several possible dispatching decisions. The objective function of the dispatching algorithm can be defined to minimise the malus of a dispatching decision. The malus can be calculated from:

- Unscheduled waiting time
- Extended running time along an alternative route
- Induced monetary costs of a dispatching decision
- Credit points of a utility function

Usually, the weight of each train class is also considered in an objective function and in the calculation of malus. To determine the impacts of a dispatching solution, the required parameters (e.g. the weights used to calculate malus) should be investigated. The parameters can be defined from many sources including experienced experts, questionnaires collected from dispatchers, or the log files recorded during daily operations.

For connection conflicts, additional waiting time may be introduced to a train to maintain a connection for another train. Hence the additional waiting time and the gain of the connections will be balanced through the calculation of malus and the evaluation of the objective function.

Today, the impacts of a dispatching solution are more or less evaluated from the point of view of railway infrastructure companies and railway operating companies. In the future, it will also be meaningful to evaluate the impacts of a dispatching solution for passengers. To define a passenger-oriented objective function, supporting data from the preferences of passengers are necessary.

The chosen dispatching strategies and the derived solutions depend on not only the objective functions, but also the time span for train dispatching as well as the dispatching territory. In addition, the possible variances that may in the future should be considered, in order to balance the robustness and the efficiency of the derived dispatching solution (Cui et al. 2012).

A given dispatching system can be either centralised or decentralised. With a centralised dispatching system, a global value of malus will be calculated for all the trains under operation. In contrast with centralised dispatching, the local malus for the involved trains are calculated in a decentralised dispatching system. In (Cui 2010), the different approaches include simulative approach, analytical approach, and heuristic approach for automatic dispatching systems are compared, and a simulation-based hybrid model for automatic train dispatching is proposed. The details of automatic train dispatching will not be covered within the scope of this book.

Timing in the prediction and resolution of conflicts plays an important role in influencing system performance and efficiency. During a simulation process, conflicts may take place frequently. Conflicts have to be predicted and resolved at each fixed time interval in a time-driven simulation, which will lead to high computational efforts. For an event-driven simulation, some events (e.g. the "departure operational point" event and the "train arrive main signal" event) are not required to activate train dispatching. The processes to predict and resolve conflicts can be arranged and controlled only at certain events, which allows for the achievement of a high level of system performance.

With the FIFO principle, conflicts are only identified at the time that a train requests its next movement authority. If the requested infrastructure resources have been allocated to other trains, an occupancy conflict is identified. However, it is too late to resolve the conflicts at the time of facing conflicts. In (Cui et al. 2017), the activity for train dispatching is not initiated at the time of requesting infrastructure resources, but at an earlier time at which a train has the chance to occupy the requested resources. Although this algorithm was originally designed for decentralised train dispatching, the same principle can also be applied for a dispatching system controlled in an OCC. The activity diagram of conflict resolution in an event-driven simulation for a centralised dispatching is shown in Figure 5-5.

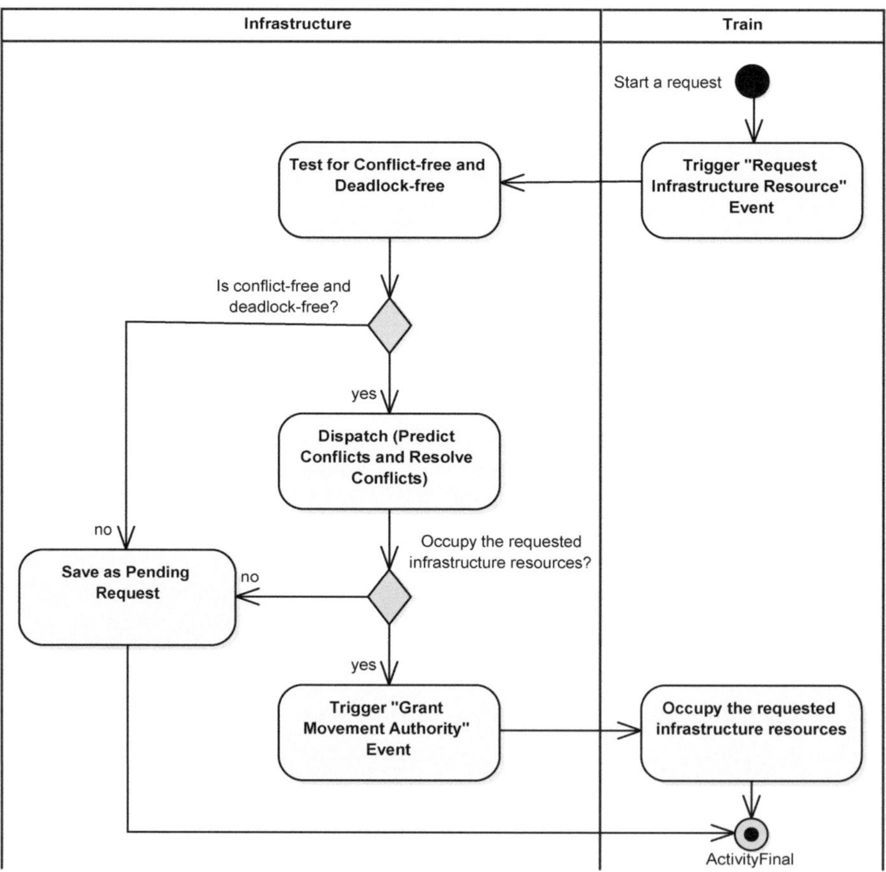

Figure 5-5: Activity diagram of conflict resolution

Once a train requests permission to occupy the next expected infrastructure resources along its path, a "request infrastructure resource" event is triggered. For centralised train dispatching, the OCC will carry out conflict-free and deadlock-free tests according to the actual operational situation. If the requested resources are not free to be occupied, or a deadlock-free situation cannot be guaranteed, the request will be saved as a **pending request** in a list, which is maintained by the OCC. For a decentralised train dispatching, each train will maintain its own pending request. The reason of the pending of the request will also be recorded, so that a new round of testing can be triggered as soon as the related operational situation has changed. The conditions to trigger a new test include the release of the involved infrastructure resources and termination of other related train runs. The conditions can be determined through the saved reason of pending. Without saving the reason, the operational situation would have to be examined frequently at a certain time interval, which may lead to high computational efforts.

In principle, a movement authority can be granted if a train has passed the conflict-free and deadlock-free tests. However, timing is also a key factor to activate dispatching system in order to predict and resolve conflicts related to the requested infrastructure resources (Cui et al. 2017). Through predicting further potential conflicts and evaluating the overall malus, the opportunity of occupancy may be given up, although the occupancy of the actual requested infrastructure resources is conflict-free and deadlock-free at a certain moment. In this case, the reason for giving up the opportunity of occupancy will be saved, and the concerned request will be placed in the list of pending requests.

The mechanism of conflict resolution can prevent the occurrence of potential conflicts and the induced hindrances at an early stage. The derived decision to give up the chance to occupy the requested infrastructure resources may significantly reduce the overall malus defined by the given objective function. Furthermore, dispatching activities are only scheduled at the time that a train has passed both the conflict-free and deadlock-free tests. A high system performance for railway simulation and train dispatching can, therefore, be ensured.

Dispatching conflicts may take place if several dispatching decisions are made locally without being coordinated through the dispatching system. The avoidance of a cyclic wait situation is of high importance, which arises if a request of a certain train is postponed in favour of another conflicting train due to a dispatching decision, the concerned train may also block the conflicting train (directly or indirectly) in proceeding with its further train run.

Figure 5-6: An example of a cyclic wait

An example of a cyclic wait is shown in Figure 5-6. For the conflict between trains T1 and T2, train T1 has been ordered to wait for train T2 by a dispatching decision. However, T2 is also not allowed to continue its train run because of a deadlock. Thus, a cyclic wait takes place. If a dispatching decision leads to a cyclic wait, the decision should be ignored. In this case, train T1 should continue its train run to occupy block section S1→S2.

A cyclic wait situation can be avoided simply through comparing the saved reason of a pending request. For example, if train T2 fails the deadlock-free test, the request will be saved as a pending request with the related reason, which includes the involved train T1 causing deadlocks. Afterwards, if train T1 is ordered to wait for train T2 by a dispatching decision, the pending status of train T2 and the reason of pending will be examined. If train T1 has already been involved in the reason of pending, the dispatching decision should be discarded. Hence, the cyclic wait will be avoided.

The cyclic wait situation due to a dispatching conflict is similar to a deadlock situation, which also leads to a circular wait (see Section 5.2). However, the reasons behind the cyclic wait are different. For a dispatching conflict, the situation of cyclic wait is due to the inconsistency of dispatching decisions, by which both train T1 and T2 are directed to wait for the other conflicting train. In a deadlock situation with circular wait, each train just continues its train run without considering other trains. In the example shown in Figure 5-6, the circular wait of a deadlock situation will take place if train T2 occupies block section S3→S4 without considering the possible conflict with train T1.

5.2 Deadlock Resolution

5.2.1 Deadlock Problem and Principles of Deadlock Resolution

In the process of a synchronous simulation, a **deadlock** situation may take place between two or more trains. In (Pachl 2011), the definition of deadlocks in railway operations are given:

"A deadlock is a self-blockade in a control system in which two or more tasks are waiting for each other to release a resource in a circular chain. In rail traffic, a deadlock is a situation in which a number of trains cannot continue their path at all because every train is blocked by another one."

Deadlock situations often take place at tracks with bidirectional operations. An example of a deadlock situation with bidirectional operations is shown in Figure 5-7. The trains T1, T2, and T3 and the paths of the trains are marked in red, green, and blue, respectively. If train T2 continues its train run to occupy block section S4→S6, all trains will be blocked by each other, resulting in the occurrence of a deadlock situation.

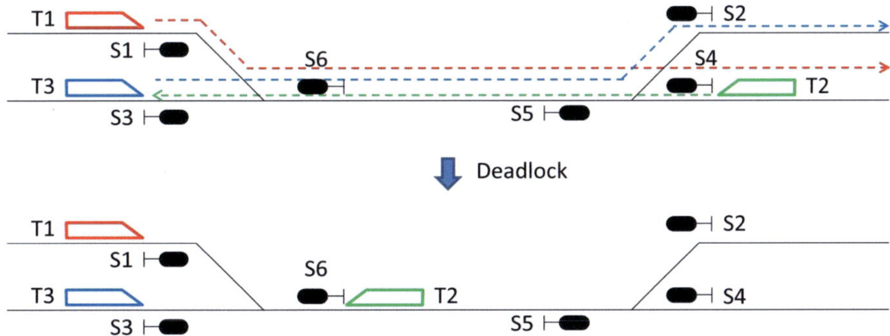

Figure 5-7: An example of a deadlock situation with bidirectional operations

In (Coffman et al. 1971), four necessary conditions to produce a deadlock situation are summarised, which are analysed for railway systems in (Pachl 2011):

    1) Mutual exclusion: a train will occupy an infrastructure resource exclusively.

2) Wait for: a train holds the occupied resources when it waits for the new requested infrastructure resources.

3) No preemption: it is impossible to remove occupied infrastructure resources from a train forcibly.

4) Circular wait: a circular wait is formed among two or more trains. Each train waits on other trains to release the required infrastructure resources. Meanwhile the train occupies the infrastructure resources requested by other trains.

The first three conditions are always in effect in railway operations. Therefore, the circular wait should be the main focus of examination for resolving deadlocks.

It is not a necessary condition that deadlocks will always occur at tracks with bidirectional operations. In railway operations, a deadlock situation may also take place at crossings at which partial route releasing has not be implemented. An example of a deadlock situation at a crossing is shown in Figure 5-8. If train T1 occupies block section S1→S2, it cannot continue its train run, since block section S2→S3 is occupied by train T2. In the meantime, train T2 is also blocked by train T1, while block section S3→S4 is not free due to the occupancy of block section S1→S2. A circular wait situation between trains T1 and T2 occurs.

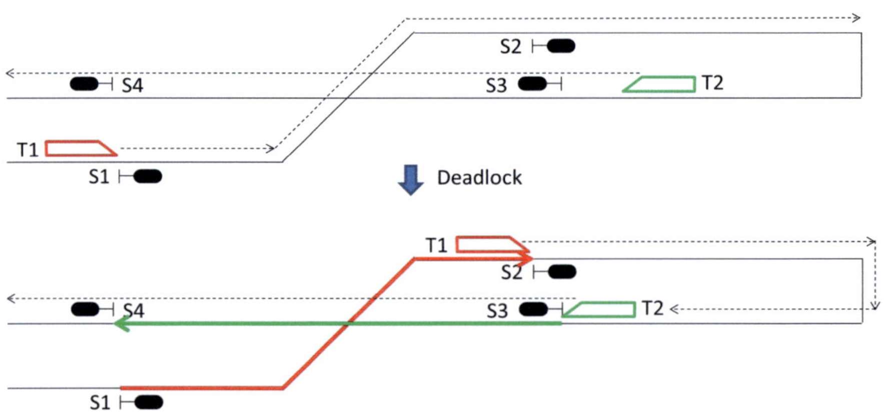

Figure 5-8: A deadlock situation at a crossing without bidirectional operations

Three possible principles to resolve deadlocks in railway operations are explained and compared in (Pachl 1993):

- Deadlock detection

- Deadlock prevention
- Deadlock avoidance

Through deadlock detection, the system in deadlock will be recovered through rolling back the previous operations. The occupied resources will be removed from the train, which leads to a deadlock situation. It does not conform to railway operations, since the condition of "no preemption" is violated. For deadlock situations in a simulation tool, it is also not efficient to frequently perform rolling back operations.

Theoretically, deadlock prevention can prevent deadlock situations through examining the four necessary conditions of deadlocks and eliminating at least one of the conditions. However, not all situations, which satisfy the four necessary conditions, will sufficiently lead to deadlocks. In most cases, it is inefficient or even impossible to apply deadlock prevention in railway operations. With deadlock prevention, the situation of circular wait should be prevented systematically, while other three necessary conditions are always in effect in railway operations. Therefore, the whole path of a train has to be reserved completely. It is impractical to apply the principle of deadlock prevention for railway operations and simulation.

The principle of deadlock avoidance, on the other hand, is suitable to resolve deadlocks for railway operations. With deadlock avoidance, the system state will be examined dynamically before an infrastructure resource is granted to a train. The request of infrastructure resources can only be approved if it is determined that the occupancy of the requested infrastructure resources will not lead to a deadlock situation. The state that can ensure a deadlock-free situation is identified as a safe state. The process to test the state for a request is called a deadlock-free test.

If the resulting state of a request is identified as a safe state, it should be guaranteed that the requested infrastructure resources could be granted to the train without deadlocks. Otherwise, a so-called "false-negative" situation happens. In contrast with "false negative" situations, a **"false-positive"** situation may occur. In a "false-positive" situation, a request, which may not lead to a deadlock, will be rejected as the result of the deadlock-free test failing to indicate a safe state. Both "false negative" and "false positive" situations should be avoided. With a "false negative" situation, a deadlock situation will take place, which indicates a failure of deadlock-test. With a "false positive" situation, the system efficiency will be limited by introducing additional waiting time. In extreme situations, a circular wait can be produced if all the involved trains are unnecessarily waiting on each other due to the "false positive" results of deadlock-free tests. In (Cui 2010), the principle of deadlock avoidance and the effects of "false positive" are discussed in detail.

5.2.2 Existing Methods of Deadlock Avoidance for Railway Systems

Since the 1980s, operations research experts have been investigated deadlock problems in transportation systems. In (Petersen, Taylor 1983), an algorithm to examine the meetability of opposing train fleets was developed. The method can identify deadlocks in networks with only single tracks and loops. However, it may frequently lead to "false positive" situations, since the trains that are not evaluated as meetable fleets through the algorithm will not sufficiently lead to deadlocks.

In (Mills, Pudney 2008), the principles of job shop scheduling were employed through a labelling algorithm. By labelling the train sequences on a single line, a feasible schedule of train runs can be derived by the vector-based algorithm. The algorithm was further developed in (Pudney, Wardrop 2008). A Kronecker algebra-based deadlock analysis was developed in (Mittermayr et al. 2012). To reduce the computational efforts, a lazy implementation is introduced with additional constraints.

The algorithms discussed above are based on pure mathematic models without integration of railway signalling and control systems. These pure mathematic models do not fit the logic for requesting, allocating and releasing railway infrastructure resources exactly. Additional assumptions and simplifications will lead to biased results that are not practical for railway operations.

In (Pachl 1993), the movement consequence analysis (MCA) method was developed to avoid circular wait through analysing the events of occupancy. The MCA can avoid deadlocks through the implementation of very strict rules. However, "false-positive" situations may take place, since it is difficult to decide how far the path of train should be evaluated in advance. The dynamic route reservation (DRR) method was further developed (Pachl 2007). The exact logic of railway operations and simulation is built within the algorithm through a route reservation model. Six rules are applied to avoid deadlocks with enforced train sequences defined by the stacked reservation model. The algorithm is under implementation by the simulation software RailSys (Pachl 2011).

The Banker's algorithm was initially applied for solving deadlock problems in railway operations and simulation in (Cui 2010). The usability and the system performance were proven by a software tool called PULRAN (Program to Research the Logistic in

Shunting Operations, in German: <u>P</u>rogramm zur <u>U</u>ntersuchung der <u>L</u>ogistik im <u>RANg</u>-ierdienst). The algorithm is further developed and evaluated in the software PULSim for large scale networks with a high density of train movements successfully.

5.2.3   The Banker's Algorithm

The Banker's algorithm (Dijkstra 1982) was originally developed to solve deadlock problems for computer operating systems. The key idea behind the algorithm is to always ensure a safe state before the requested resources (e.g. money for banking systems, infrastructure resources in railway systems) are granted. A safe state means that all the requested resources can satisfy the needs of all the requesters (e.g. customers of a banking system, or trains in a railway system). If a safe state cannot be guaranteed, the requester should wait until other requesters return their allocated resources, at which point a safe state can be reached again. The banker's algorithm is categorised as an algorithm of deadlock avoidance (see Section 5.2.1), by which a deadlock-free test should be carried out before allocating resources.

In the Banker's algorithm, the occupancy of resources and the requests of resources should be known in advance in order to perform deadlock-free tests. The preconditions of the Banker's algorithm fit railway operations and simulation very well. The required information can be obtained from the occupancy situation of infrastructure resources and the scheduled train runs. In (Cui 2010), the Banker's algorithm is introduced for railway simulations. Three sets of resources are defined:

- Currently available resources (AR): the available infrastructure resources that have not been occupied by any train yet.
- Currently occupied resources (OR): the infrastructure resources occupied by a train.
- Maximal required resources (MR): all infrastructure resources needed by the trains in their further train runs.

Once a train (a requester) requests to occupy one or more infrastructure resources, a deadlock-free test starts. Within the deadlock-free test, it is assumed that the train will occupy the requested infrastructure resources. Therefore, the requested infrastructure resources belong to the set of OR instead of MR. A set of trains P, which contains all the trains that are satisfied with the needs of their requested resources, will be initialised as empty at the beginning. All the trains will be tested with several rounds. In each round, all trains that are not in the set P will be checked. If the requested infrastructure resources of a train are in the AR or OR, the passed train will be put into the set P. All the previously occupied resources of the passed train will be returned to AR. The tests are carried out iteratively with several rounds until all the trains are in the set P or no more trains can be added into the set P by the current round of tests. In the latter situation, the requests of the rest trains are not able to be satisfied furthermore, in which case the state will be deemed unsafe. In case all the trains are in the set P, a safe state is ensured. The requested infrastructure resources can be therefore granted to the trains without deadlocks. The activity diagram of deadlock-free test is shown in Figure 5-9:

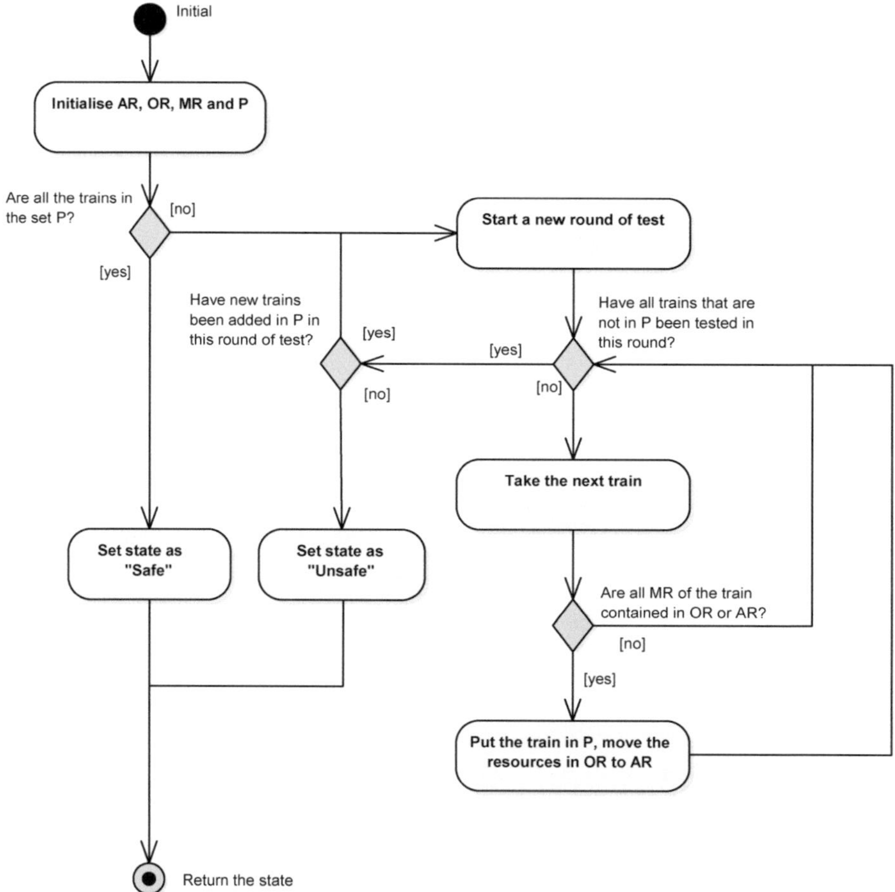

Figure 5-9: Activity diagram of the Banker's algorithm

In (Cui 2010), infrastructure resources are modelled in the form of infrastructure elements such as tracks, turn outs, and crossings (see Section 2.2.4), which are suitable for the network with non-signalised operations (e.g. shunting movements) or for moving block systems. In the software PULSim, infrastructure resources are modelled as one or more path components for signalised operations. A train will apply all the path components of the requested block section. In the following paragraphs, the deadlock-free tests for a signalised network in Figure 5-10 are carried out for trains T2 (marked in green) and T3 (marked in blue). In this example, the infrastructure resources are requested in the form of block sections instead of path components for comprehensive understanding without excessive unnecessary details.

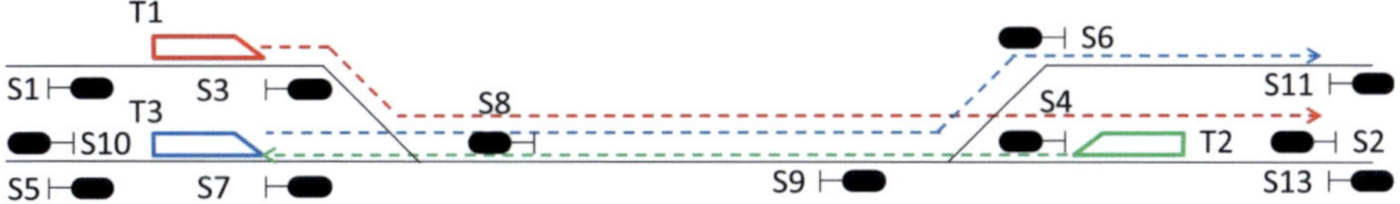

Figure 5-10: A signalised network for deadlock-free tests with the Banker's algorithm

Train T2 is requesting the infrastructure resource S4→S8. The request should be rejected since it will lead to an unsafe state when subjected to the deadlock-free test. The procedure of the deadlock-free test is shown in Table 5-1:

Round: Activity	Request of S4→S8 for the train T2
Initialise	P: {}, AR: {}
Round 1: test for T1	OR: {S1→S3}, MR: {S3→S9, S9→S13} Not all the resources in MR are in OR or AR.
test for T2	OR: {S2→S4, S4→S8}, MR: {S8→S10} Not all the resources in MR are in OR or AR.
test for T3	OR: {S5→S7}, MR: {S7→S9, S9→S11} Not all the resources in MR are in OR or AR.
Round 1: finished	No new trains have been added in P
	Unsafe state

Table 5-1: Deadlock-free test for the train T2

Train T3 is requesting the infrastructure resource S7→S9. The process of the deadlock-free test is shown in Table 5-2:

Round: Activity	Request of S7→S9 for the train T3
Initialise	P: {}, AR: {S9→S11}
Round 1: test for T1	OR: {S1→S3}, MR: {S3→S9, S9→S13} Not all the resources in MR are in OR or AR.
test for T2	OR: {S2→S4}, MR: {S4→S8, S8→S10} Not all the resources in MR are in OR or AR.
test for T3	OR: {S5→S7, S7→S9}, MR: {S9→S11} All the resources in MR are in OR or AR. Put T3 in P, return all OR to AR
Round 2:	A new train T3 has been added in P. P: {T3}, AR: {S5→S7, S7→S9, S9→S11, S4→S8, S8→S10}
test for T1	OR: {S1→S3}, MR: {S3→S9, S9→S13} Not all the resources in MR are in OR or AR.
test for T2	OR: {S2→S4}, MR: {S4→S8, S8→S10} All the resources in MR are in OR or AR. Put T2 in P, return all OR to AR
Round 3:	A new train T2 has been added in P. P: {T2, T3}, AR: {S5→S7, S7→S9, S9→S11, S4→S8, S8→S10, S2→S4, S3→S9, S9→S13}
test for T1	OR: {S1→S3}, MR: {S3→S9, S9→S13} All the resources in MR are in OR or AR. Put T1 in P, return all OR to AR. All trains in the set P: {T1, T2, T3}.
	Safe state

Table 5-2: Deadlock-free test for the train T3

The infrastructure resource S7→S9 can be granted to train T3, since the deadlock-free test leads to a safe state. For deadlock-free tests with signalised operations, if an infrastructure resource (a path component or a block section) is returned to the set AR, all the conflicting block sections, which were not free due to the occupancy of the returned infrastructure resources, should also be returned if they are free. For example, at the end of round 1, block sections S5→S7 and S7→S9 are returned to AR. The conflicting block sections S4→S8 and S8→S10 are also returned to AR at this moment.

If a request has passed the deadlock-free test with the Banker's algorithm, the requested infrastructure resources can be granted in a safe state. It sufficiently guarantees that deadlocks will not take place. Therefore, the "false-negative" situations will never take place with the Banker's algorithm. However, the "false-positive" situation can still occur. The false-positive situations and the improvements of the Banker's algorithm will be discussed in Section 5.2.4.

5.2.4   False-Positive Situations and Improvements of the Banker's Algorithm

Passing the test through the Banker's algorithm is a sufficient condition but not a necessary condition for deadlock avoidance. Some requests, which may not actually lead to deadlocks, will still be rejected due to a failed deadlock-free test through the Banker's algorithm. This situation is called "false-positive". A false-positive situation will produce additional waiting time unnecessarily and will therefore result in low utilisation level of infrastructure resources.

Figure 5-11: An example of false-positive situations

An example of a false-positive situation and the corresponding improvement is shown in Figure 5-11. The tested train T1 requests to occupy block section S1→S2. The request will be rejected through Banker's algorithm, since train T2 occupies the requested resource of T1 and vice versa. However, once T1 reaches the position before signal S2 (the position of T1 is marked with a dashed line), the result of the deadlock-free test will not be presented through Banker's algorithm. At this situation, train T1 does not block the requested infrastructure resource of T2 anymore. A false-positive situation of the Banker's algorithm results due to a lack of consideration of potential future state transitions.

In (Cui 2010), four improvements of Banker's algorithm are presented:

- A: analysis with potential state transitions
- B: test processes in a right order
- C: test for alternative routes
- D: test only for the request with non-junction-type resources

These improvements have been tested and proven for small networks in the work of (Cui 2010). In recent years, the usability and the system performance have been further examined and improved through real examples in large networks with huge amounts of train movements.

With improvement B (test processes in a right order), the trains that will leave the investigated area and the trains with an internal destination lying at a dead-end track will be tested firstly. Therefore, those trains will free the occupied resources before the destination of their paths, and the unnecessary blockage of other trains will be avoided. Using improvement D, only those requests with non-junction-type resources are tested, since the occupancy of junction-type resources will not change the state for testing deadlocks. Both improvement B and improvement D should be applied as standard features in the test with Banker's algorithm to improve the system efficiency. Improvement C examines the possibilities of using alternative routes. It depends highly on the applied dispatching algorithm as it is not possible to use any available alternative routes without knowing the strategy of the dispatching system. However, the algorithm for testing deadlocks should be independent of the dispatching system in most cases. In addition, the test with alternative routes leads to high computational efforts, since the full tests through the Banker's algorithm will be carried out completely for several alternative routes. Therefore, the improvement with alternative routes is not applicable as a standard feature for the deadlock-free test.

In (Cui 2010), improvement A is implemented through new rounds of tests with the Banker's algorithm. It increases the computational efforts considerably for large scale networks. A new, more-efficient algorithm of using feasible resources is therefore developed in Section 5.2.5.

5.2.5 Algorithm for Testing Deadlocks with Feasible Resources

The idea behind the algorithm for using feasible resources is to consider possible state transitions from an unsafe state to a safe state due to the change of the position of the tested train in its further train run. **Feasible resources** are one or more connected infrastructure resources in the further path of the tested train. The feasible resources have not been granted to the tested train yet. If feasible resources are available, the infrastructure from the requested resources to the feasible resources can be granted to the tested train, even the train has not passed the deadlock-free test through the Banker's algorithm.

In (Cui 2010), the term "feasible tracks" was defined for the improvement of analysis with potential state transitions. Here the concept of feasible resources is further developed based on the concept of feasible tracks. Instead of testing deadlocks through the Banker's algorithm once again, the feasible resources are determined through examining conflict situations with other trains. Feasible resources should satisfy the following criteria:

Criterion 1): All the resources in the path from the current position of the train to the feasible resources are free to enter.

Criterion 2): The feasible resources should be long enough to hold the tested train completely.

Criterion 3): If the tested train has occupied the feasible resources, the other trains should not merge, cross or oppose with the tested train at the feasible resources.

Criterion 4): There is no other train merging with the tested train at the infrastructure resource from the current position until the feasible resources.

At first, the train must be free to enter the resources from the current position to the feasible resources. This is guaranteed by Criterion 1. If feasible resources are found, the block sections from the current position to the feasible resources will be blocked. Criterion 2 should be satisfied in order to ensure the feasible resources are long enough to hold the tested train. Meanwhile the other infrastructure resources occupied by the tested train can be released. Hence, the unsafe state caused by occupancy of the requested infrastructure resources is avoided.

To fulfil the Criterion 3, the occupancy situation between the tested train and the other trains will be checked. If other trains merge, cross or oppose with the tested train at the feasible resources, it may lead to deadlocks.

In Figure 5-12 a), the situation of merging is illustrated. Block section S1→S2 is checked for the tested train T1, which is merging with train T2. If T1 occupies block section S1→S2 (T1 is marked with dashed lines), T2 is not able to continue its train run. Hence a deadlock situation takes place among T1, T2, and T3. Block section S1→S2 is not a feasible resource, and is not allowed to be occupied by T1. The situation of crossing is shown in Figure 5-12 b). If T1 occupies the block section S1→S2, train T2 has to wait in front of signal S4. Therefore, a deadlock situation takes place. Similarly, in Figure 5-12 c), train T1 should wait in front of signal S1. Otherwise, trains T2 and T1 fall into a deadlock situation. Block section S1→S2 is not a feasible resource for the tested train T1.

In this section, the possible deadlock situations are demonstrated in a "small" infrastructure network with a small number of trains. In reality, many involved trains can form a large loop of deadlocks in a large area.

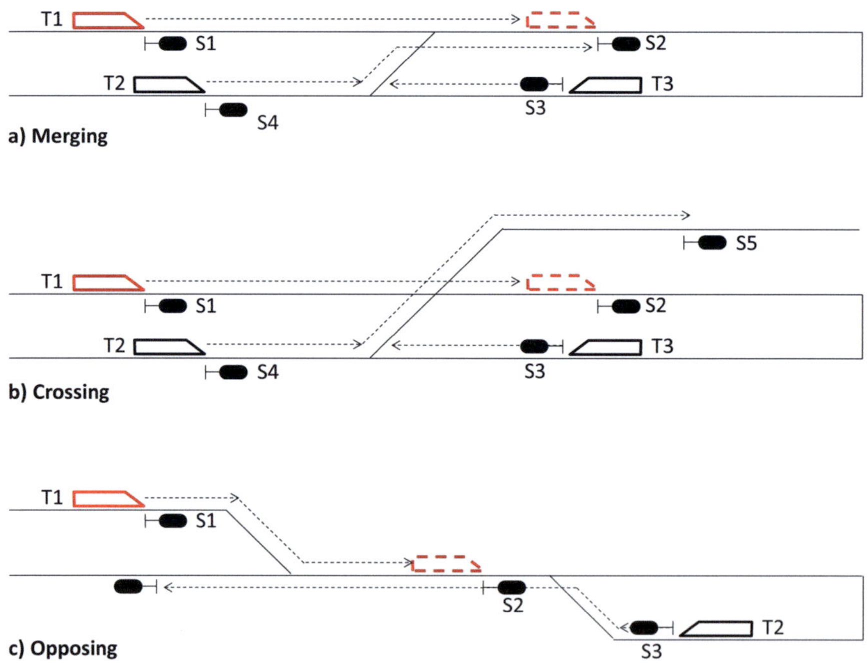

Figure 5-12: Deadlock situations of testing feasible resources for merging, crossing and opposing with criterion 3

Criterion 4 should be satisfied in order to avoid the tested train being blocked by a merged train, which can produce a deadlock situation due to the blockage of the merged train. In Figure 5-13, trains T2 and T1 are merging at the infrastructure resource before the checked feasible resource S2→S3. Block section S2→S3 is not allowed to be granted to the tested train T1 as a feasible resource. Otherwise, train T2 would have to wait at signal S4 or S2, which would lead to deadlocks among trains T1, T2, and T3.

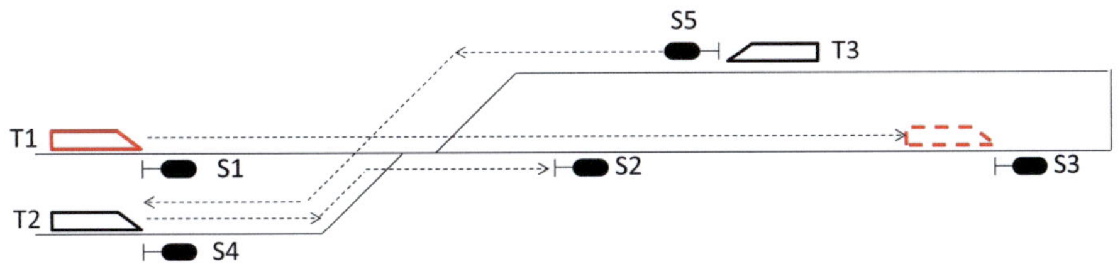

Figure 5-13: A deadlock situation of merging at an infrastructure resource before a checked feasible resource for criterion 4

The feasible resources can be determined based on the four aforementioned criteria. The examination of the feasible resources will be carried out along the path of the tested train. Once feasible resources are found, the tested train can continue its train run to occupy the feasible resources, at which a deadlock-free situation can be guaranteed.

As defined in Section 2.4.5, an infrastructure resource is the basic unit for requesting, allocating and releasing. For a signalised network with partial route releasing, several infrastructure resources (path components) are allocated together as a block section. After the train has passed a partial releasing point, the infrastructure resources will be released one by one.

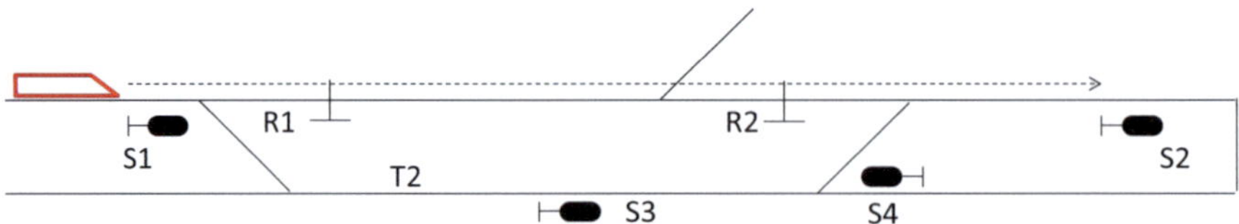

Figure 5-14: Allocating and partial releasing a block section

In Figure 5-14, an example of allocating and partially releasing a block section is illustrated. Three infrastructure resources, S1→R1, R1→R2, and R2→S2, are allocated at the same time for block section S1→S2. After the train leaves the partial releasing point R1, the infrastructure resource S1→R1 will be released, while other infrastructure resources (R1→R2 and R2→S2) are still occupied by the train. After the train leaves the partial releasing point R2, the infrastructure resource R1→R2 will be released, and the infrastructure resource R2→S2 will be occupied by the train.

Because of the different timings for allocating and releasing infrastructure resources, the examination of feasible resources for a block section should be carried out for all the non-released infrastructure resources (path components) of that block section. The non-released infrastructure resources include the currently examined path component and the subsequent path components until the end of the block section.

In Figure 5-15, the examination of feasible resources for block section S1→S2 will be started for path components S1→R1, R1→R2, and R2→S2. If these block sections do not satisfy the four criteria, path components R1→R2 and R2→S2 will be checked for the feasibility of occupying R1→R2 and R2→S2 (path component S1→R1 may be released). In this example, even though R1→R2 satisfies the four criteria; it cannot be granted to the tested train T2 as a feasible resource, since a merging situation takes place between trains T2 and T1 at R2→S2. If the tested train T1 occupies R1→R2, path component R2→S2 will also be blocked by the tested train T1. Hence a deadlock among T1, T2, and T3 takes place.

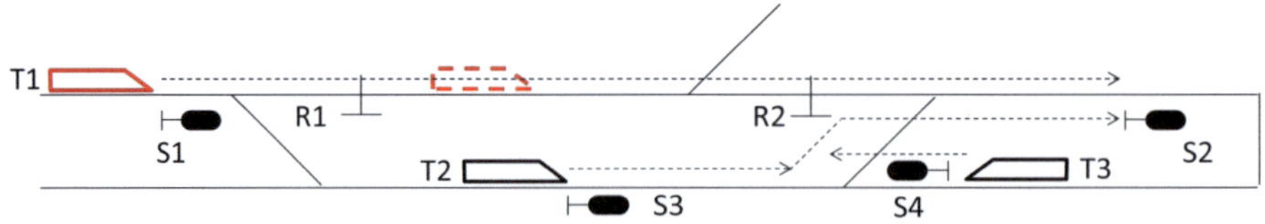

Figure 5-15: Examining feasible resources for all the not-released path components

The workflow of examining feasible tracks is illustrated in Figure 5-16. The objective in searching for feasible resources is to eliminate circular wait situations. It can be conceived as an approach of deadlock prevention (see Section 5.2.1). The combination of deadlock avoidance (with the Banker's algorithm) and deadlock prevention (through searching for feasible resources) can efficiently decrease the waiting time caused by false-positive situations with limited computational efforts and simple logics. Since the developed method using deadlock prevention is only applied from the current position of the tested train to the feasible resources, the inflexibility of being blocked by the whole line with deadlock prevention can be avoided. In Section 5.3, the advantages of using feasible resources to avoid false-positive situations are presented.

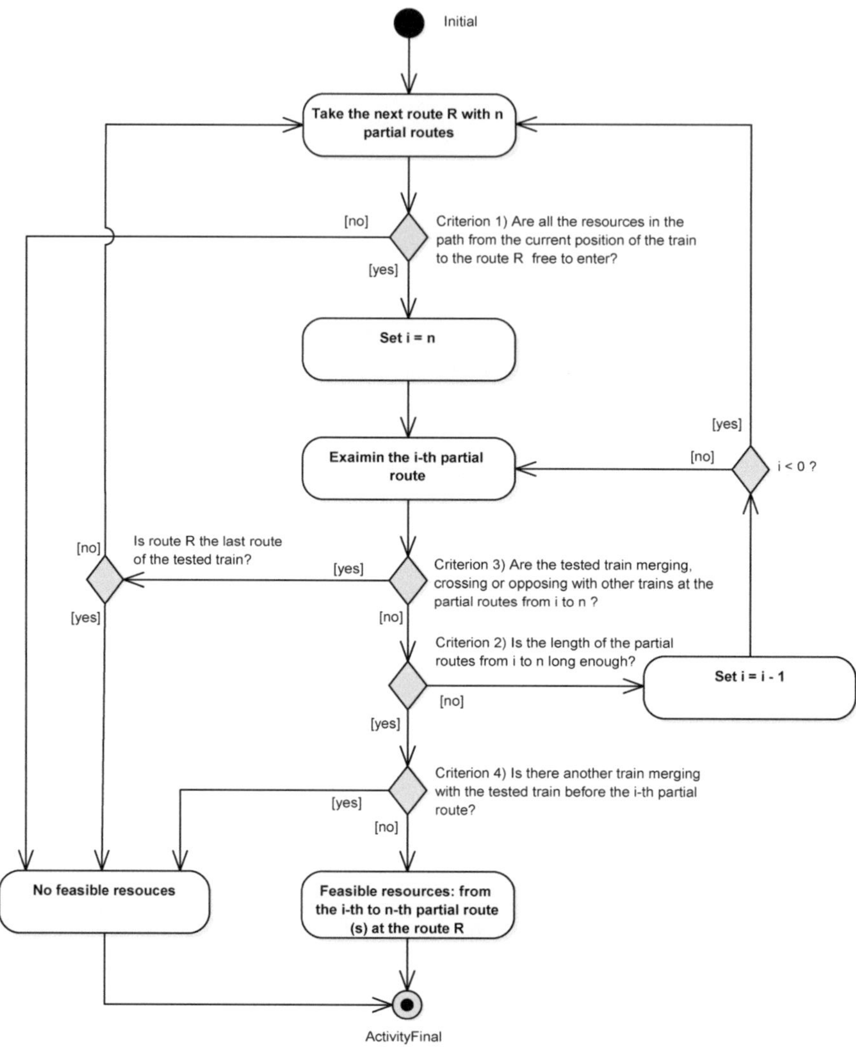

Figure 5-16: Work flow of examining feasible resources

5.3 Practical Applications with the Banker's Algorithm

The Banker's algorithm with identification of feasible resources for deadlock avoidance has been applied and tested in the simulation software PULSim for two different case studies.

In the first case (Case 1), the investigated network has 71 stations with 1,350 train runs in one day. The deadlock-free tests have been carried out for the whole day a total of 11,580 times, wherein 11,214 tests have been passed with the Banker's algorithm. There are also 356 requests of infrastructure resources that have been approved after the feasible resources were identified. Although these 356 requests have not passed the test through the Banker's algorithm, the train can still be ensured to be in a safe state at the feasible resources. There are ten requests in total that have not passed the deadlock-free tests.

The algorithm is further tested and evaluated for another case (Case 2) in a network with 129 stations and 2388 train runs in one day. In this example, there exists a terminal main station and a high number of opposing train movements at several locations. Therefore, the chance of the existence of deadlock situations is very high. Within this case, 36,254 tests have been carried out in total, wherein 27,026 tests passed through the Banker's algorithm. Furthermore, 8,057 tests have been approved due to the identification of feasible resources. Hence the waiting time caused by the false-positive situation can be avoided. For Case 2, there are 3.23% tests in total that failed through the testing of deadlocks, and the rate of reduced false-positive situations is 22.22%.

The average execution time for deadlock-free tests in Case 1 is less than 1 millisecond, while the average execution time of deadlock-free tests for Case 2 is around 3.5 milliseconds. The reason behind the increased execution time in Case 2 is the large scale of the network, as well as the efforts required to identify the feasible resources. In Case 1, most of deadlock-free tests were passed through the Banker's algorithm. Only 366 requests in Case 1 were further examined for feasible resources. In Case 2, 9,228 requests were failed by the Banker's algorithm. Therefore, additional execution time is required to test the availability of feasible resources for these 9,228 requests.

In both Case 1 and Case 2, it remains to be determined how far the test of feasible resources should be carried out. In Case 2, the farthest feasible resource has been found in the next 11-th block section. If the feasible resource in the next 11-th block section is used to avoid deadlocks, the whole path of the next ten block sections will be locked and reserved for the train, which will result in a high waiting time and hindrance for other trains. Therefore, it is practical to set an integer value for searching depth to limit the number of block sections for further search, upon which the search for feasible resources will be terminated. Several case studies are carried out for case 2, in order to compare the performance with different settings of depth. In Figure 5-17, the reduced rate of false-positive situations with different values of searching depth is shown. For these case studies, the rate of reduced false-positive situations remains stable after the depth for searching for feasible resources is set as three. It is still not clear, whether an "optimal" value of the upper limit exists for all cases or not. The setting of the upper limit should be investigated in further research.

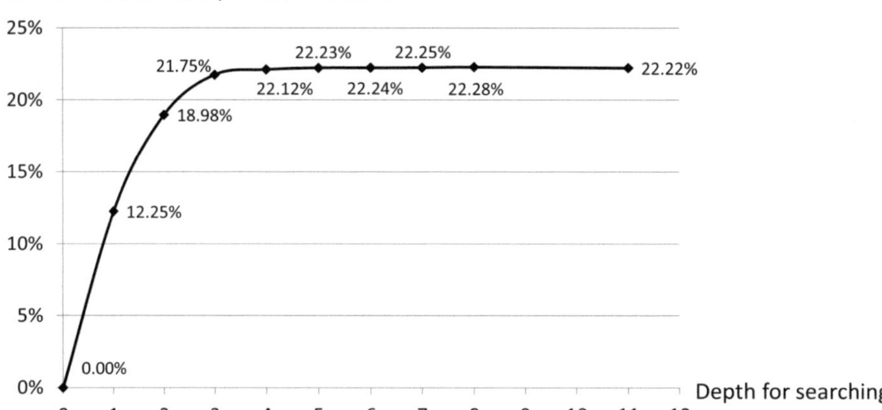

Figure 5-17: Rate of reduced false-positive situations for different setting of depth in Case 2

6 Calibration of Railway Simulation Models

6.1 Objectives of Calibration for Railway Simulation

It is always a challenge to determine whether a simulation model can represent the real world accurately. For railway simulation, the elimination of all the discrepancies between simulated results and real values cannot be guaranteed. The discrepancies of a simulation model are caused by:

- Modelling errors: Due to the insufficient knowledge or the simplification of reality, the simulation models are not able to represent the railway system exactly.
- Errors of input data: Incomplete and/or inaccurate input data and parameters will produce biased simulation results.
- Errors of applied parameters: The applied parameters influence the results of the calculation/simulation. The errors of the applied parameters can lead to discrepancies.

Modelling errors are usually presented in the form of systematic errors. For example, if some special resistances are not considered during a running time calculation, the simulated running time may be smaller than the value in reality. The modelling error can be reduced through the improvement of the algorithm and the model of simulation software.

One of the most common modelling errors is the incorrect implementation of simulation software. If there are some programming bugs, the model will lead to biased results. Therefore, it is important to carry out systematic tests to ensure the correctness of the simulation software.

The term "**verification**" refers to the process of determining whether a model has been implemented correctly in computer software. Several testing techniques will be applied for verification. The key idea of software testing is to compare the simulation results with the expected value, which is usually derived from a theoretical calculation. For example, in a running time calculation with the velocity-step approach, the train movements at each velocity step can be calculated for many discrete points based on Newton's law (see Section 4.1.1). The simulation result can be verified by comparing it with the calculated running dynamics. It is impossible to verify all the possible executions of a simulation tool. A comprehensive discussion of the technique for software testing is introduced in (Kaner et al. 1999).

The verification for railway simulation must be executed not only for the model and the software implementation, but also the input data of infrastructure, rolling stocks, and operations. During the verification process, the level of detail should be taken into consideration (see Section 2.1.1). The accuracy of the data depends of the purpose of simulation. For example, highly abstracted and aggregated information is sufficient for a macroscopic model, which can be used for a general, preliminary planning. The transport time and the headway of a line are often of concern for a macroscopic simulation, and the exact blocking time is not required. However, the data used for a macroscopic simulation is not sufficiently detailed for timetabling with a high requirement of accuracy. If the exact data of infrastructure and rolling stocks is not available, the running time, the blocking time and the conflicts cannot be calculated correctly. A simulation model is always built based on certain simplifications and assumptions. The purpose of simulation, the efficiency of the software, and the level of details of the model should be carefully balanced.

In timetable simulation or operational simulation (see Section 3.3), the parameters used in the simulation model may also lead to discrepancies. For example, the parameters used to model train resistance (see Section 2.3.2) influence the accuracy of the results in a running time calculation. The parameters used for a running time calculation are the most fundamental parameters in timetable simulation. In operational simulation, the stochastic processes are modelled through the parameters that define the distribution of disturbances. One of the most important tasks of operational simulation is to determine the disturbance parameters, in order to derive the service quality of the investigated operating program.

To prove the credibility and the correctness of a simulation model, the simulation results should be examined. The term "**validation**" refers to the process of comparing the simulated results and the data collection in reality. A simulation model is the simplification and the approximation of reality, and therefore, there is no absolutely valid model. A certain criteria of validation should be set in advance based on the purpose of the simulation. If necessary, the parameters should be further corrected in order to minimise discrepancies. The process of validating and correcting the parameters used in a simulation model is called **calibration**. In this chapter, the basic algorithm for calibration of parameters will be presented.

Several different calibration approaches are presented in Section 6.2. The calibrated parameters can be used for timetable (deterministic) simulation or (stochastic) operational simulation, which will be discussed in Sections 6.3 and 6.4, respectively.

During the calibration process, the data of reality are required to be used as benchmarks. The availability of the data of reality determines the difficulty of calibration (Law 2007). For example, if the quality of railway services of a planned operating program that is supposed to be used in future will be evaluated through an operational simulation, the resulting delays for the planned operating program in reality are not available yet. It is impossible to calibrate the model through comparing the simulated data and the data of reality directly. Therefore, the model and the used parameters will be calibrated for comparable existing systems as references. The assumption behind the approach is that the same relationship between the stochastic disturbances and the resulting delays can be applied for both the existing system and the planned system. It is a qualified assumption for the situation if no existing system is available. The method for calibrating the parameters of an operational simulation will be discussed in Section 6.4.

6.2 Approaches for Calibration of Simulation Models

In a calibration process, the term "**indicator**" refers to the data collected in reality. The simulated results will be compared with the indicators in order to verify them and to calibrate the parameters used in a simulation model.

The collected real data can be applied directly as indicators in a calibration process. For example, the calculated running time of a simulation model can be verified with the logged data directly. In this paper, the directly used indicators are called original indicators. Usually the original indicators are used for calibrating parameters in a deterministic process (e.g. in a timetable simulation).

It is of no value to compare the stochastic value of real data directly with the simulated results. In an operational simulation, the statistical value of delays (e.g. the mean value of delays) is of greater concern than an individual delay at one stop for one single train run. Therefore, the statistical indicators are applied for calibrating parameters in a stochastic process (e.g. in an operational simulation).

A calibration process can be carried out manually by experienced experts. This method is suitable in a small investigation area for a limited number of parameters with a strong relationship between the calibrated parameters and the indicators. In this situation, the experienced experts can find the right direction to correct the value of parameters efficiently. However, the complexity and the efforts of calibration will be increased dramatically for a large scale network with a high density of train runs. Hence an automatic calibration system is desired.

A practical approach is to calibrate parameters based on regression analysis, which is a statistical process used to estimate the relationships between the indicators and the parameters. The principles and the techniques of regression analysis are presented in (Draper, Smith 1998). Through observing and comparing the changes in the indicators due to the varied value of the parameters, the model between the dependent variables (the indicators) and the independent variables (the parameters) can be established. For regression analysis, a mathematical equation to model the relation between the studied variables (the parameters and the indicators) should be available. The mathematical equation is constructed according to certain assumptions. For example, linear regression is a very common approach to model a linear prediction function. The assumption of using linear regression is that the relationship between the calibrated parameter and the indicators should be linear. In railway timetabling, the regular recovery time is modelled in the form of a percentage of the minimum running time. For an existing timetable with a calculated minimum running time and a given scheduled running time, the percentage for calculating regular recovery time can be calibrated through linear regression (Zhao et al. 2016).

The approach of regression analysis is widely used to analyse and to model railway systems. In (Kecman, Goverde 2015), the data-driven approaches with Least Trimmed Squares (LTS) robust linear regression, regression trees, and random forests are applied to model running time and dwell time for railway operations. The methods of using regression analysis for railway simulation is also discussed in (Sipilä 2015).

In recent years, the approaches with machine learning have been paid more attention. Machine learning can be categorised as supervised learning, unsupervised learning, semi-supervised learning, reinforcement learning or deep learning. There are many common characteristics between regression analysis and machine learning. For example, the approach of supervised learning is used to infer a function derived from training data, so that new examples can be produced from the inferred function. The training data for supervised learning are the indicators used in a calibration system. The mathematical function built in a regression analysis can be perceived as the inferred function for supervised learning. Indeed, both regression analysis and machine learning have learned the methodology from each other. The two approaches have become closer in recent decades.

In contrast with regression analysis, in which explicit mathematical equations are required, a statistic model in the form of mathematical equations is not necessarily presented in machine learning. The objectives of machine learning are to identify the patterns of input data (e.g. from unsupervised learning) and to estimate the possible outputs. Instead of using an equation to calculate the outputs, the results are derived and estimated from an inferred model built through machine learning. It is not the target to reveal and explain the relationship between the studied samples and outputs with a mathematical equation directly.

In (Markovića et al. 2015), a machine learning model with support vector regression is developed to estimate the influences of infrastructure on arrival delays. An approach using machine learning for calibrating parameters in operational simulation will be presented in Section 6.4 in detail.

6.3 Calibration of Timetable Simulation

The credibility of a railway simulation tool highly depends on the quality of its calibration for timetable simulation. The most fundamental tasks of timetable simulation are to construct timetables and to identify conflicts. The other applications of timetable simulation, including calculation of waiting time and synchronisation time, determination of capacity, and identification of bottlenecks, can be carried out based on the derived running dynamics and conflicts of train runs.

As presented in Section 4.1, the minimum running time of a train run can be calculated based on the given data regarding infrastructure and rolling stocks. The calculated minimum running time may have discrepancies when compared to the real value. However, it is impossible to know the real data of the minimum running time. The calculation of the minimum running time should be examined through a verification process. Based on the given data, the simulated results will be compared with the value derived from the analytical calculation.

Sometimes the existing timetable can be used to validate the minimum running time. The assumption of the validation method is that the existing timetables are constructed realistically, and that the recovery time and the waiting time are known. Through this approach, the simulated running time can be kept in consistence with the given timetable. A method of validation of the minimum running time based on the given timetable is developed in (Zhao et al. 2016).

In reality, the real running time of a train run can be larger than the minimum running time. In a simulated model, the additional running time between two reference points should be distributed along the two reference points. Depending on the design and the implementation of a simulation tool, the additional running time can be determined in different ways. Usually, the scheduled running time should be taken into consideration so that the punctuality of the train run can be ensured. In the software PULSim, the current delay of a train run is also used to calculate the additional running time. The calculation of additional running time between two reference points in PULSim is:

$$T_{additional} = T_{scheduled} - T_{minimum} - T_{delay} \qquad (6\text{-}1)$$

Notations used:

$T_{additional}$ Additional running time between two reference points [seconds]

$T_{scheduled}$ Scheduled running time between two reference points [seconds]

$T_{minimum}$ Calculated minimum running time between two reference points [seconds]

T_{delay} Delay of the current train run [seconds]

Here the value of T_{delay} can be positive, zero or negative. If the delay of the current train run is positive, the train will try to run faster to catch up to the scheduled timetable. If the delay of the current train run is negative, the train will run slowly to "consume" the extra running time. This approach is suitable for calculating running dynamics for an individual train run. If multiple train runs are considered, the driving style will also be influenced by dispatching systems.

With a determined additional running time between two reference points, it is still unclear how the additional running time is distributed along the train run in a simulation model. Some examples of the distribution of the additional running time can be:

- To distribute the additional running time as early as possible
- To distribute the additional running time as late as possible
- To distribute the additional running time equally along the train run

- To distribute the additional running time with optimised minimum energy consumption (used in PULSim, see Appendix B)

The distribution of the additional running time depends on the driving style of the train as well as the operational situation. Based on the collected data of running dynamics during railway operations, the parameters to simulate the train run should be calibrated. For example, GPS data can be collected to calibrate train movements (Medeossi et al. 2011). In (Fabris et al. 2008), the recorded on-board train events are analysed to improve the accuracy of microscopic simulation models.

After the running dynamics of a train run has been derived, the blocking time stairway can be constructed in a simulation model. The parameters of the signalling system are decisive to improve the accuracy of the simulation model. For example, if the value of the time required to build a route is too small, the blocking time for a block section will be shorter than the real value in reality. Hence some potential conflicts may be missed in the simulation model.

Usually the exact data of blocking time in railway operations are not available. The blocking time and the parameters of signalling system can be calibrated through the logged conflicts. For example, the train describer log files with infrastructure and train description messages are available in the Dutch train describer system TNV and TROTS (Kecman, Goverde 2012). In (Daamen et al. 2008) and (Goverde et al. 2008), the route conflicts can be identified automatically, which can be used to analyse the consequences of route conflicts and the consecutive delays. These data can be used for further calibration of the construction of blocking time stairways.

The algorithm used for calibrating the parameters of timetable simulation depends on specific simulation tools, the available reality-based data, and the required level of detail and the purpose of the simulation. Since the process of timetable simulation is deterministic, the calibration is often based on the original indicators collected in reality. The availability and the quality of the real data are the key success factors of timetable calibration.

6.4 Calibration of Stochastic Disturbances in Operational Simulation

For an operational simulation, it is decisive to generate realistic disturbances sufficiently close to the reality. The disturbances are modelled through a set of statistical parameters, which are called disturbance parameters (see Section 3.3). The objective of calibration of stochastic disturbances in operational simulation is to determine the value of these disturbance parameters.

In (Cui et al. 2014) and (Cui et al. 2016), an algorithm using machine learning to automatically calibrate the disturbance parameters was developed. In the calibration algorithm, the mean value of delays and their probabilities are used as statistical indicators for the comparison of the simulated delays and the delays measured and collected from the real situation.

During the calibration process, the value of the disturbance parameters will be adjusted iteratively. After each round of iteration, the simulated delays will be used to calculate the instantaneous value of indicators (the mean value of delays and the probability of delays). The instantaneous value of indicators will be compared to the target value of the indicators, which are measured and collected from reality. Guided by the calibration algorithm, the disturbance parameters will be updated continuously to minimise the difference between the target value and the instantaneous value of indicators until their convergence. For disturbance parameters θ, the resulting difference between the target value and the instantaneous value $J(\theta)$ is called cost. The cost function used to calculate cost is defined in (Cui et al. 2016):

$$J(\theta^{(t)}) = \frac{1}{2 \cdot q \cdot m} \sum_{i=1}^{m} \sum_{j=1}^{q} (v_{i,j}^{\theta^{(t)}} - \overline{v_{i,j}})^2 \qquad (6\text{-}2)$$

Notations used:

$\theta^{(t)}$ The disturbance parameters in the t -th iteration, [-]

$J(\theta^{(t)})$ The value of the cost with disturbance parameters $\theta^{(t)}$, [-]

$v_{i,j}^{\theta^{(t)}}$ The instantaneous value for the j-th indicator of delays for the measurement point i with disturbance parameters $\theta^{(t)}$, [-]

$\overline{v_{i,j}}$ The target value for the j-th indicator of delays for the measurement point i, [-]

m The number of the measurement points, [-]

q The number of the considered indicators for a delay, [-]

The gradient descent algorithm is used to update the value of disturbance parameters and to minimise costs. After the $(t-1)$-th round of the iteration, the value of an individual parameter θ_i at the t-th iteration will be updated as:

$$\theta_i^{(t)} = \theta_i^{(t-1)} - \alpha \cdot \frac{\partial J(\theta)}{\partial \theta_i} \qquad (6\text{-}3)$$

Notations used:

$\theta_i^{(t)}$ The value of the disturbance parameter θ_i in the t-th iteration, [-]

$\theta_i^{(t-1)}$ The value of the disturbance parameter θ_i in the $(t-1)$-th iteration, [-]

α The learning rate, which controls the step length of each iteration, [-]

$\frac{\partial J(\theta)}{\partial \theta_i}$ The partial derivative of $J(\theta)$ with respect to the disturbance parameter θ_i, [-]

The idea behind the gradient descent algorithm is to adjust the parameter towards the minimum value of the cost. The sign of the partial derivative indicates the change of the cost caused by the varied value of the parameter. In Figure 6-1, the different situations of updating the parameters are illustrated. At point A with a negative partial derivative, the updated value of $\theta_i^{(t)}$ is increased from the value of $\theta_i^{(t-1)}$. It results that the value of θ_i is adjusted towards the direction of the minimum value of cost. Similarly, the value of the parameter will be decreased at point B with a positive partial derivative. The setting of the learning rate α can be adjusted dynamically with the Resilient Backpropagation (RPROP)-Algorithm (Riedmiller, Braun 1993) to improve system performance. In (Cui et al. 2014) and (Cui et al. 2016), the method of establishing learning rate and calculating the partial derivatives are presented in detail.

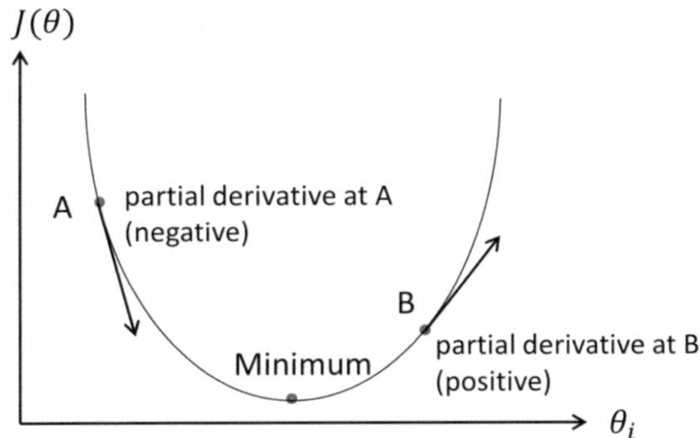

Figure 6-1: An example of the gradient descent algorithm

It is difficult or even impossible to build a mathematic model for the relationship between disturbance parameters and indicators, which is the main reason for the use of the machine learning approach. During the calibration process, it is not the aim to build a precise statistical model to describe the relationship among the disturbances, the primary delays and the consecutive delays. Instead, the influences on the delays from the various disturbance parameters will be learnt, in order to determine the calibrated parameters with the best fit between the simulated delays and the actual delays.

The developed calibration algorithm has been tested and evaluated in accordance with several different scenarios. The results show that an automatic calibration can improve the efficiency and the accuracy of an operational simulation. For the reference railway network used in (Cui et al. 2014), consisting of four stations with 10 train runs per hour, an external simulation software was applied. The error of the mean value of delays between the simulated results and the target value was reduced from 11.54 seconds (manual calibration) to 4.17 seconds (automatic calibration). Additionally, the error of the probability of delays was reduced from 1.82% (manual calibration) to 0.92% (automatic calibration). For the same quality requirements of calibration (the same errors of the mean values and the probability of delays between the simulated results and the target value), the time efforts have been reduced from 8.5 hours (manual calibration) to 2.6 hours (automatic calibration). If a reasonable initial value of disturbance parameters is set by a skilled user, the time efforts for the automatic calibration can be further reduced (this would also be the case for manual calibration).

Integrated with the software PULSim, the system performance for calibration was further improved (Cui et al. 2016), and the frequent disk I/O (input/output) operations with external simulation software were avoided. For the investigated network with 71 stations and 72 train runs per hour, the total calibration time is 5.1 hours with 64 rounds of iterations. In (Cui et al. 2016), the published results of the calibration processes are presented.

In further development, several techniques of machine learning should be applied in order to improve the quality and the accuracy of the calibration system. The gradient descent algorithm is used to calibrate the parameters and ensures a good level of conformity of the most influencing parameters. However, it cannot guarantee a globally optimal solution. To overcome this problem, several rounds of calibration can be carried out with randomly generated initial values. It is also worthwhile to apply an algorithm to estimate the value of the parameters. The estimated values can be used as the initial values of parameters for further calibration. With the derived initial values, the convergence can be speed up, and the calibrated parameters can be also verified with the estimated values.

The unsupervised learning will be then applied to identify the indicators, which present significant biases among the others. The calibration process can be carried out in several steps. In each step, the value of the influencing parameters calibrated in previous steps will be fixed, and the other less-influencing parameters will be further calibrated. The efforts of an individual calibration step with the gradient descent algorithm can be reduced with a limit on the number of parameters. However, the efforts of the entire calibration process with several steps should be balanced. Again, the design of the calibration system depends on the purpose of the investigation and the required level of detail.

7 Evaluation of Railway Operations and Interfaces of Railway Simulation Tools

7.1 Evaluation of Railway Operations

The simulation of railway operations serves as a starting point for evaluating and improving the quality of railway services. For a given infrastructure network and operating program, the simulated results will be further processed and analysed. The tools for evaluation and optimisation can be either directly integrated within a simulation software, or be developed separately through the published interfaces provided by a simulation software. For example, a software package called "evaluation manager" is available in RailSys as a built-in evaluation tool (RMCON 2010). The function of evaluation can also be implemented by a third party. For example, based on the simulation results generated by RailSys, the software tool PULEIV (Program to Capacity Research, in German: Programm zur Untersuchung des Leistungsverhaltens) has been further developed by the IEV, Universität Stuttgart for the purposes of capacity research and evaluation of the quality of railway services (Martin et al. 2011).

The evaluation of simulation data can be carried out online or offline. The comparison of online evaluation and offline evaluation is shown in Figure 7-1.

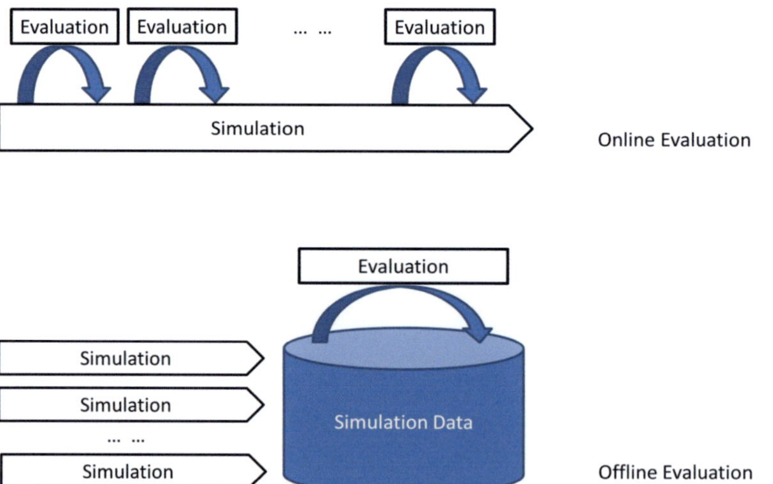

Figure 7-1: Comparison of online evaluation and offline evaluation

An online evaluation is possible for a built-in evaluation package or a simulation tool with open interfaces for online data access (see Section 7.3). Online evaluation is required for real-time tasks such as train dispatching, by which the conflicts should be identified and evaluated in time to derive a dispatching decision. Online evaluation is usually very efficient, since the time-consuming I/O operations can be forgone through the built-in simulation model or the open interfaces.

For a middle term or a long term planning and evaluation, the computational efforts are often considerable with a very large amount of data being generated during simulations. An offline evaluation will be applied to process the massive data in such cases. Usually, for an offline evaluation, it is only possible to analyse and evaluate the results after several rounds of simulation. For example, in capacity research, various operating programs should be simulated initially. The quality of services for each operating program will be compared and evaluated afterwards. Another example can be seen in the process for calibrating disturbance parameters (see Section 6.4). The effects of adjusting the disturbance parameters are evaluated through calculating the value of the indicators (delays) after a round of operational simulation.

Depending on the purpose of the research, online and offline evaluation can also be combined. For example, to investigate the influences of dispatching on the capacity of railway systems, the relationship between the dispatching parameters and the resulting delays can be analysed online. Meanwhile, the influences on the capacity of the railway system will be evaluated offline.

Several approaches, including simulative approaches and analytical approaches, are available to evaluate the quality of railway services. Currently, simulative approaches are also often used to examine and to improve analytical models. The statistical indicators (see (Schwanhäußer 1978), (UIC 2013), (Vakhtel 2002), (DB Netz AG 2008), and (Martin et al. 2014)) used in both simulative and analytical approaches can be derived from the simulation data.

The most important simulation data that can be directly obtained during the execution of a simulation are:

- Running dynamics of each train run in the form of discrete points (see Section 4.1.1)
- Blocking time stairway of each train run
- Time and the position for requesting, applying and releasing infrastructure resources for each train run

- Delay of a train at a stop or a measurement point
- Identified conflicts and deadlocks
- Decisions made by the dispatching system

For a time-driven simulation, the process and the interaction among infrastructure and trains at each time interval can be recorded for a detailed analysis of train runs. Similarly, in an event-driven simulation the events are recorded for a detailed evaluation of the railway operations.

The occupancy of infrastructure and the hindrances during train running are of special concern. These values can be used for further investigations such as capacity research and train dispatching. The calculation of the occupancy time and the hindrances will be discussed in Section 7.2.

The evaluation of a given simulation can be carried out at different levels of detail. For a macroscopic evaluation, the aggregated statistical information such as the average delays and the capacity of an investigated network are studied. The aggregated macroscopic indicators can be further observed at a mesoscopic level. For example, the operating performance for a certain group of trains at a station, a junction-type basic structure or a station tracks can be investigated for a mesoscopic evaluation. Since a macroscopic or a mesoscopic evaluation can be aggregated from the most detailed information on a microscopic level, the calculation of the occupancy of infrastructure and the hindrance of trains presented in Section 7.2 focuses on the detailed microscopic level.

7.2 Occupancy of Infrastructure and Hindrance of Trains

7.2.1 Calculation of Occupancy Time

The exploitation of infrastructure can be measured through the occupancy time of each infrastructure resource in the investigated time period. The occupancy rate is also used in some cases, which is the value of the occupancy time divided by the time period of investigation.

The allocating and releasing times of each train run can be obtained directly from the respective recorded blocking times of the train run. An infrastructure resource for a signalised network can be a block section or a path component (with partial route releasing). In the case of railway operations with moving blocks or in a non-signalised network, an infrastructure resource can consist of a set of directed edges of tracks or junction-type elements (see Section 2.4.5). Since infrastructure resources can lead to overlapping, it is not intuitive to demonstrate the occupancy of infrastructure based on infrastructure resources. In a railway network, the infrastructure elements (see Section 2.2.4.) can be categorised as junction-type infrastructure elements (e.g. turn-outs, crossings) and non-junction-type infrastructure elements (tracks). The occupancy of the infrastructure will be calculated for junction-type infrastructure elements and non-junction-type infrastructure elements, respectively.

A junction-type infrastructure element will always be requested, allocated and re-leased as a whole. Therefore, the occupancy time of a junction-type infrastructure element can be calculated through summing up the blocking time of the infrastructure element for all the train runs. This approach can be applied for the signalling systems with fixed blocks or moving blocks.

For a non-junction-type infrastructure element (track), the occupancy of the track will be divided into several segments with a fixed-distance block system. For each seg-ment, the occupancy time of the infrastructure resources from different directions will be added. In (Martin et al. 2014), this approach is used to calculate the occupancy time for basic structures and path components (see Section 2.2.6).

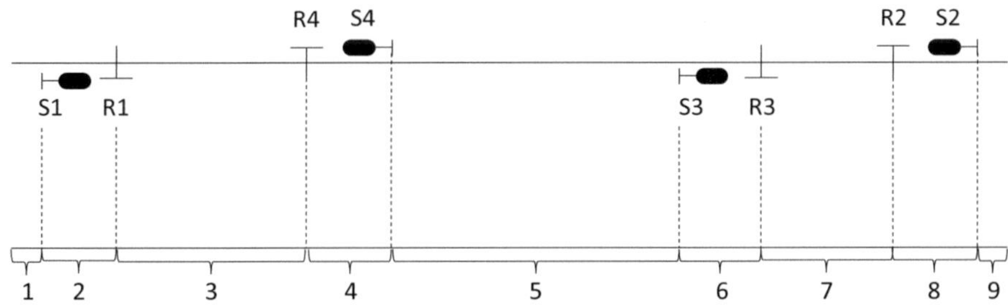

Figure 7-2: An example of separating a track for the calculation of occupancy time

An example of separating a track for calculation of the occupancy time for a fixed block system is shown in Figure 7-2. There are two different running directions at the track. The signals and the releasing points of route/signals are used for separating the track. Nine occupancy segments (marked with a series of numbers, respectively) divide the track, taking the signals, the releasing points, and the starting/ending point of the track as boundaries. Inside each occupancy segment, the occupancy time (see Section 7.2.2) should be the same. In Figure 7-3, an example of occupancy time in a station is illustrated. The duration of the occupancy time at each infrastructure element is represented by the thickness of the green lines.

Figure 7-3: An example of occupancy time in a station

To calculate the occupancy time of infrastructure with moving blocks, the blocking time of each train run at a track should be initially determined. In Figure 7-4, the method of determining blocking time of a train run at a track is illustrated. With the calculated running dynamics of the discrete points (e.g. based on the velocity-step method) for a train run, the occupancy time at each discrete point can be determined (see Section 4.2.3). The blocking time of two adjacent discrete points is derived by connecting the endpoints of the occupancy time at each discrete point. In Figure 7-4, the blocking time of the two adjacent discrete points (point i-1 and point i) is marked in the area with light blue colour.

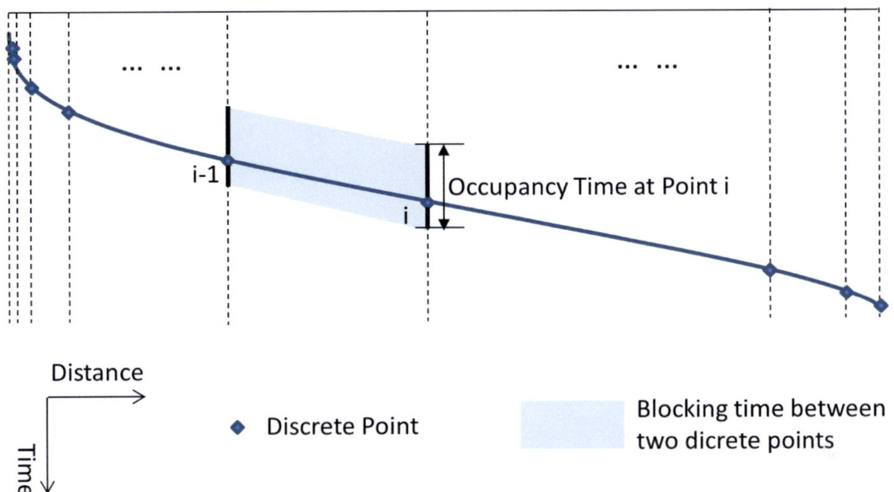

Figure 7-4: Determination of blocking time of a train run at tracks for moving block system

After the blocking time of each train run has been determined, the occupancy time of a track can be calculated. If the velocity-step approach is applied, the length between two adjacent discrete points (e.g. point i-1 and point i) for an individual train run can be variable. On a track, the occupancy time for a certain time period is calculated for several train runs. The division of the track though discrete points for each train run is different. Therefore, it is meaningful to divide a track equally into many small segments with fixed lengths. The occupancy time of a train run at each segment will be interpolated from the blocking time for the middle point of the segment. For each segment, the occupancy time can be calculated through summing up the interpolated occupancy times for all the train runs. In Figure 7-5, the occupancy time of segment S for two trains runs is the sum of the interpolated occupancy time (marked with the red colour) at the middle point of segment S.

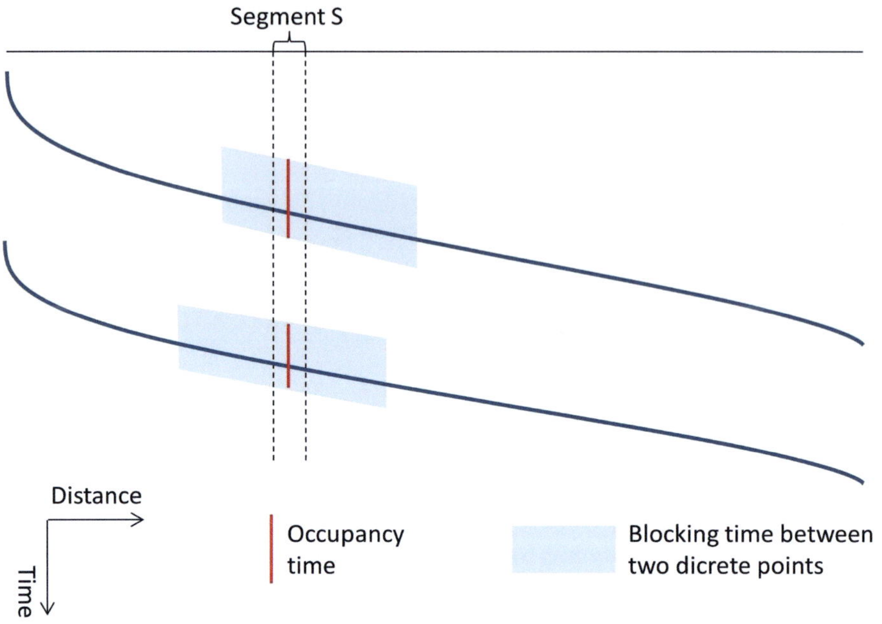

Figure 7-5: Calculation of occupancy time for a segment of a track in moving block system

7.2.2   Hindrances of Trains: the Pending time and the Waiting Time

In order to evaluate the quality of railway services, the hindrances of trains are used to measure the influences among train runs. An important objective of improving the quality of railway services is to minimise the hindrances of trains.

It is very common to measure hindrances of trains through the calculation of waiting times. **Waiting time** is defined as the extension of running time and dwell time due to hindrances from other trains (DB Netz AG 2008). Waiting time can be categorised as unscheduled waiting time and scheduled waiting time. Unscheduled waiting time is known as the delay, which is caused by the stochastic disturbances during railway operations. Scheduled waiting time is the artificial increase in the overall time of a train caused by the resolution of conflicts during the scheduling process (Hansen, Pachl 2014).

For a macroscopic evaluation, the unscheduled waiting time is directly collected from the delays. The delay at a stop or a measurement point is measured through comparing the difference between the actual and the scheduled departure/arrival/passing time. Delays are classified as primary delays (including original delays and initial delay) and consecutive delays (Hansen, Pachl 2014). The measured delays can be primary delays, consecutive delays or the mixture of both. In (Goverde 2005), a max-plus model was applied to analyse timetable stability and delay propagation. The behaviour of delay propagation was formalised with an activity graph in (Büker, Seybold 2012). It is very practical to use the convolution of the distributions to model the interaction among stochastic events. In (Yuan 2006), an approach using a numerical method for calculating the convolution of delay distribution was presented. In capacity research, the outgoing delays are compared with incoming delays for a investigated network in order to assess the ability of the investigated network to recover from delays (Schmidt 2009).

Usually, the scheduled waiting time at the macroscopic level is not directly recorded in a timetable. To derive the scheduled waiting time, a simulation of an individual train run without consideration of other train runs (see Chapter 4) is first carried out and recorded. The scheduled waiting time can then be determined through comparison of the given timetable and the information for the case without hindrance, recorded by the simulation of an individual train run. At the macroscopic level, the scheduled waiting time is only measured at a stop or a measurement point.

At the microscopic level, the waiting time will be measured for each point where a hindrance takes place. The scheduled waiting time is measured through comparing the result of the timetable simulation and the simulation result of an individual train run without hindrances. The unscheduled waiting time should be measured through comparison of the result of the timetable simulation (without disturbances and dispatching) and the result of each round of an operational simulation. Since the scheduled waiting time has been already included in the result of the timetable simulation, the resulting unscheduled waiting time describes the impacts of the hindrances caused by the random disturbances.

It is not always straightforward to measure the hindrances through waiting time at the microscopic level. Once another train hinders a train, the resulting waiting time may not always be positive. For a hindrance with a short time period, the waiting time can be compensated from the recovery time. The recovery time refers to the regular recovery time and the special recovery time, which are included in regular running time. In a timetable, the scheduled running time includes regular running time, scheduled waiting time and scheduled synchronisation time. The recovery time will be derived through comparing the regular running time and the minimum running time.

A negative waiting time can sometimes result, despite a train being hindered while applying to the next infrastructure resource. The negative waiting time is due to the unevenly distributed recovery time from various driving behaviours. For a situation without hindrances, the velocity of the train may be slowed down if the available re-covery time at further infrastructure resources is large enough. Once the train is hin-dered, the recovery time for the yet unallocated infrastructure resource is unknown. The train may run faster than the case without the hindrance before it stops at the current EOA. In the case that the requested infrastructure resource is available be-fore the train starts braking, the train will receive the movement authority of the re-quested infrastructure resource. It will run under the situation without hindrances again. In this situation, the resulting waiting time will be negative, since the running time will be shortened at a high velocity.

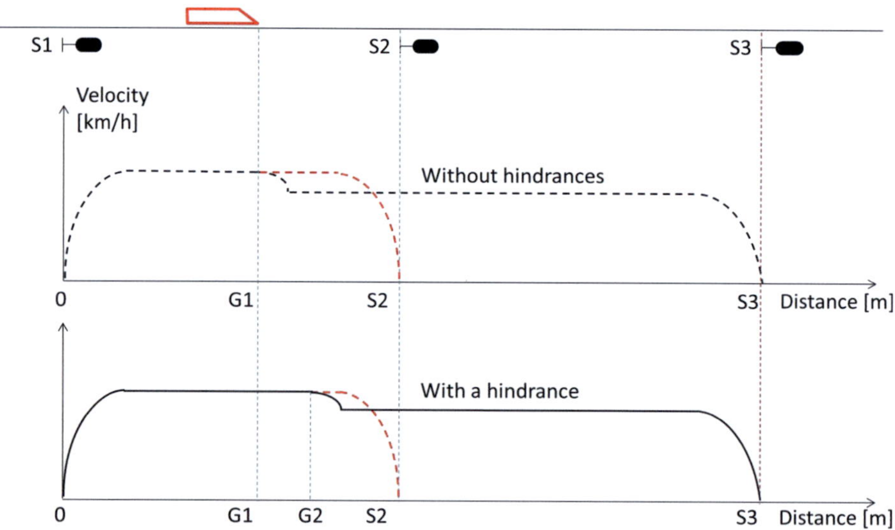

Figure 7-6: An example of negative waiting time

An example of negative waiting time is shown in Figure 7-6. At point G1, if the train can obtain the movement authority for the next block section S2→S3 without hin-drances, the recovery time at block section S2→S3 will then be known. The velocity of the train will be slowed down after point G1 due to the reserves. In case of hindrances, the block section S2→S3 will not be available for the train at point G1. The train still maintains a high velocity until G2, where block section S2→S3 can be granted. Between G1 and G2, the train runs faster than the situation without hindrances. Hence it results a negative waiting time.

For a hindered train, the negative impacts of hindrances (e.g. unscheduled braking and re-acceleration) cannot be identified if the value of waiting time is zero or negative. Therefore, the **pending time** of a request can be used to evaluate the impacts of hindrances. The pending time is defined as the time period from the time at which the requested infrastructure is supposed to be allocated to the train in case of no hindrances, to the actual time that the train obtains the movement authority to enter the requested infrastructure resource.

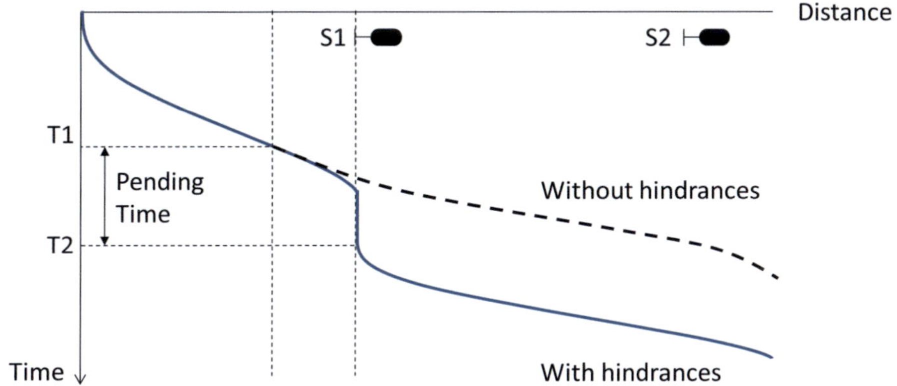

Figure 7-7: Example of pending time

Pending time can be obtained during the process of simulation, an example of which is illustrated in Figure 7-7. The original train path without hindrances is shown with a dashed line. At time point $T1$, the actual train path diverges due to the hindrance for requesting the block section S1→S2. When the movement authority is granted to the train at time point $T2$, the pending time will be derived as $T2 - T1$. The pending time can be perceived as the waiting time before the requested infrastructure is available. For any hindrance, the pending time will always be able to be identified as a positive value. Moreover, the relationship between the pending time and the overlapping of blocking time from two conflicting trains is intuitive. Therefore, the pending time of a request can be used to evaluate the quality of railway services and to predict the impacts of conflicts with a high level of accuracy. An example of pending time in a station is shown in Figure 7-8. Similar to the example shown in Figure 7-3, the duration of the pending time is represented by the thickness of the red line.

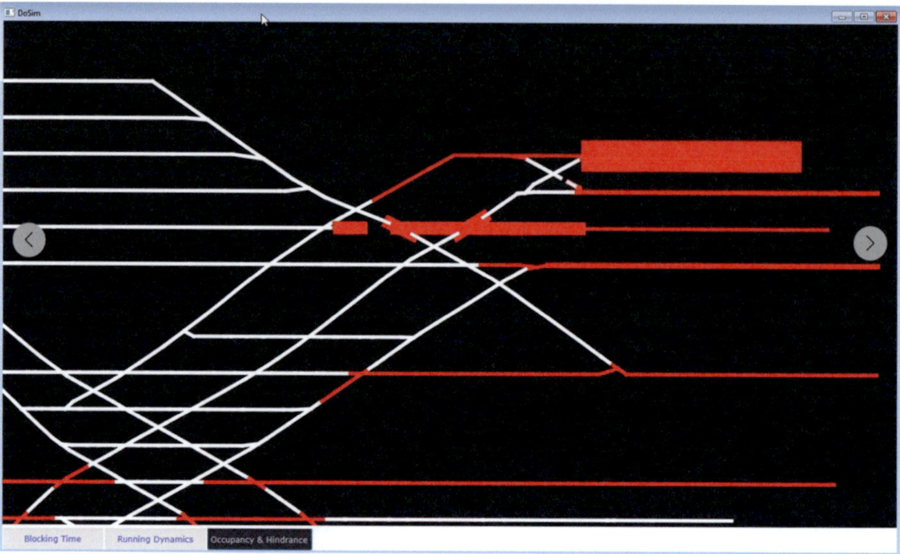

Figure 7-8: An example of pending time in a station

7.3 Open Interface of Railway Simulation Tools

7.3.1 Log Files of Railway Simulation for Offline evaluation

It is very common to use log files for evaluating the results of a simulation. The log files are usually stored as XML files, CSV files or plain text files. The log files can be read and analysed by integrated evaluation packages or third party software (if the data format of the log files is published). The log files are often used for offline evaluation (see Section 7.1).

Today, an open and standardised data format to describe the results of railway simulation is not yet available. Although a standardised data format is desirable, it is difficult to unify different simulation tools to meet the wide variety of requirements and concerns of the users. Depending on the purpose of simulation, certain kinds of data formats are defined by each individual simulation tool. For example, the published data formats for RailSys are (RMCON 2010):

- Single simulation: the information of each train run, including the start and end station, the number of the current end signal, the line, the current velocity, the delay, and the hindrance, are logged in this file.
- Multiple simulations: For an operational simulation, the overtaking and dispatching activities, deadlocks, the most significant delays and hindrances, as well as other statistic information, are logged in this file.
- Dispatching: the dispatching measures and the resulting delays are logged in this file for further evaluation.
- Running time calculation: the minimum running time and the scheduled running time are listed and compared in this file. The actual running time will be calibrated according to the scheduled running time.
- Events during train run: depending on the system setting, this file records the events of train run at a station or a signal.
- Utilisation of block sections: The occupancy time and delay propagation on each block are recorded in this file.

Since the I/O operation is very time-consuming, the user can decide the settings of the log files in order to improve system performance and to reduce disk usage. Although flexibility and system performance are limited in the evaluation process using log files, the saved log files can be reused in the future for other purposes of evaluation without needing to run a simulation again.

At times, unpublished log files are available for certain simulation tools. They are mainly used for internal testing for some special purposes. Although the information can be inferred from the contents of the log files, it is not recommended to use these data due to the lack of documentation and support.

7.3.2   Evaluation from Open Application Programming Interface (API)

The results of a simulation can be retrieved dynamically through an open Application Programming Interface (API), which enables an online evaluation. If the functions of the evaluation are integrated with the simulation software, the simulation output can be directly called upon by the software package for evaluation. For a third party application, an open API will provide access from the evaluation tool to the simulation tool. With the open API, the process can be simulated, analysed, and optimised in real-time.

An API based on web services is available in the simulation tool OpenTrack (Seybold, Huerlimann 2016). Based on web services, the third party tool can send messages as commands to OpenTrack, and retrieve the results in the form of status messages back to the tool (Figure 7-9). The messages are warped according to the Simple Object Access Protocol (SOAP) over Hypertext Transfer Protocol (HTTP).

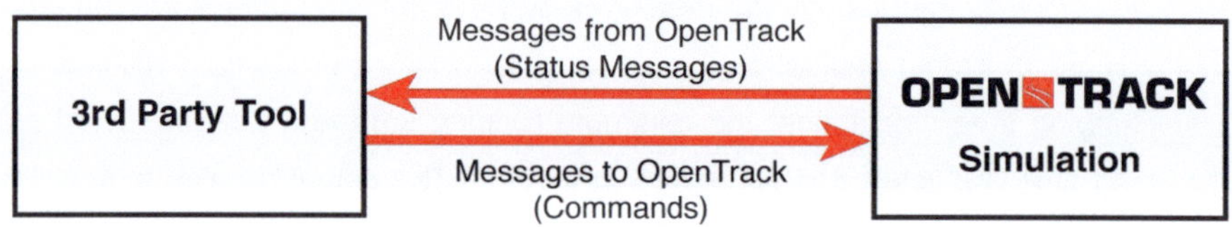

Figure 7-9: OpenTrack API (Seybold, Huerlimann 2016)

With an open API, a simulation tool is open to be accessed by third party applications. It is possible to integrate the simulation tool within the processes of evaluation and optimisation seamlessly. In (Seybold, Huerlimann 2016), an application of using OpenTrack API for train dispatching is presented. Served as a testing environment, the simulation tool OpenTrack replaces reality via exchanging the commands and status messages in the same format of information used in reality. Hence new dispatching algorithms and prototypes can be tested in a simulation environment.

The trade-off of the integration is the increased efforts of communication and implementation. For example, the solution of web service with SOAP can impact the system performance during enveloping and transporting the messages via XML data. Especially for a large amount of data, e.g. infrastructure data or the detailed results of a running time calculation, the overhead of communication is considerably high. The development of web service also depends on large software vendors and tools. It is not feasible for researchers and railway engineers who are interested in lightweight solutions to evaluate and optimise the quality of railway services.

7.3.3 Open Interfaces with Scripting Languages

A scripting language is a programming language optimised in the form of a series of imperative commands, which can be executed in a certain run-time environment.

During the execution of a script language, the code is usually interpreted within the run-time environment without needing to be compiled. Therefore, a script language can be used to customise a sophisticated process with many tasks and interactions. A script language usually hides the internal structure and the implementation details of a complex system. It is suitable to be provided as a shell of a complex system, so that the user can only concentrate on the integration of the functions of the system. Hence the complexity of learning the whole system and the efforts for further development are significantly reduced. Through the ability to quickly learn and write the commands in an imperative way, end users can achieve their programming objectives flexibly and efficiently.

Scripting languages can be categorised as general-purpose languages and domain-specific languages. Some examples of well-known and popular general-purpose scripting languages are Python, Perl, and Ruby. The general-purpose scripting languages can be used in various applications with plenty of support and documentation. Domain-specific languages are specially developed for a certain platform and a single application. An external interpreter/parser for the domain-specific language is needed.

An example of a domain-specific language is SimTalk, which is used in the simulation software Plant Simulation (Bangsow 2010). Plant Simulation belongs to Siemens PLM software, which is a software suite for product lifecycle management and manufacturing operations management. SimTalk is applied as a scripting language for the software Plant Simulation. The language of SimTalk is divided into a section of control structures, including conditions and loops, as well as a section for built-in standard methods to control objects (called "materials" in Plant Simulation) and information flows.

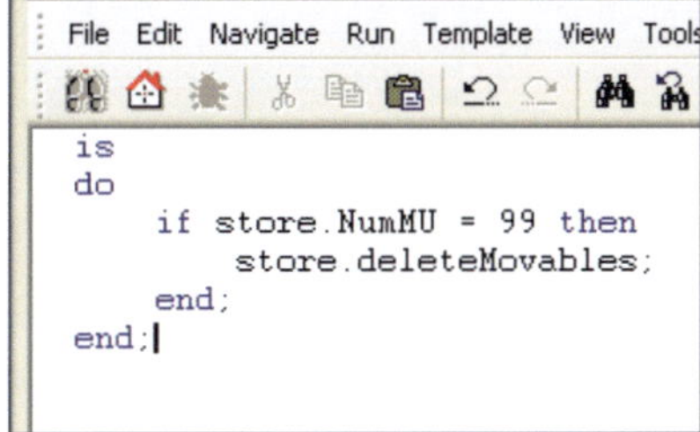


```
is
do
        if store.NumMU = 99 then
              store.deleteMovables;
        end;
end;
```

Figure 7-10: An example of screenshot of the SimTalk editor (Bangsow 2010)

In Figure 7-10, a screenshot of the SimTalk editor is illustrated. The "if-then" and the "do-end" clauses control the workflow of the procedure. The attribute and method of the object "store" are used to read and change the state of the "store" object. The methods of the object serve as commands in SimTalk. Users can interact with the system through the execution of the script following the control structures.

A script language for railway simulation is not yet available. In the software PULRAN (Cui 2009), an XML-based file is developed to model the railway system and the movements of shunting processes. Users can define the desired movements of trains either through editing the XML file or through interacting with a graphical user interface. The movements of trains are defined as "actions", which are similar to the commands used in a scripting language. The defined actions will be saved in the XML file. An example of the defined actions is shown in Figure 7-11. A train with the Id "Z1-0" is supposed to run to the destination at the track "Gleis8". The distance between the starting port of the track and the target position is 61.218 meters.

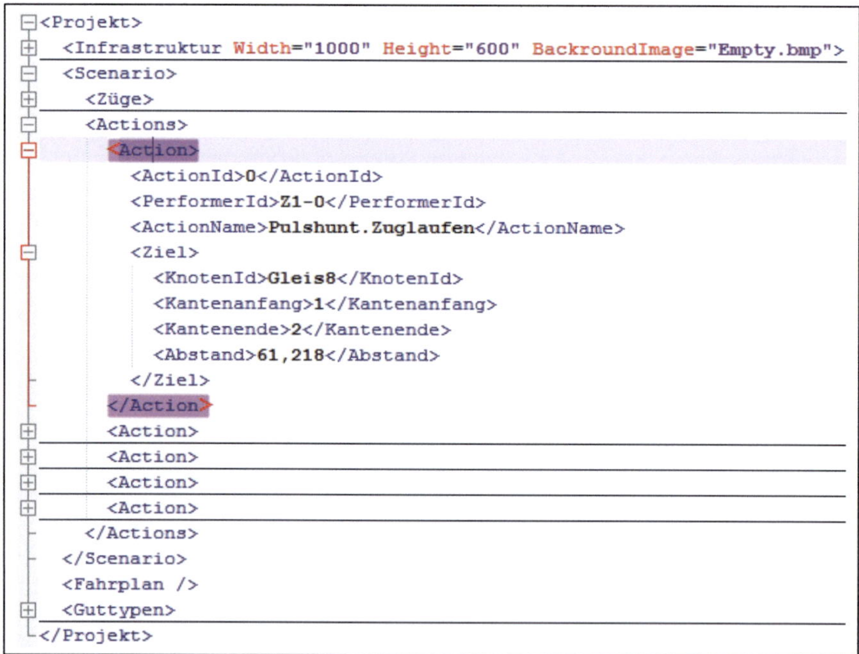

```
<Projekt>
  <Infrastruktur Width="1000" Height="600" BackroundImage="Empty.bmp">
  <Scenario>
    <Züge>
    <Actions>
      <Action>
        <ActionId>0</ActionId>
        <PerformerId>Z1-0</PerformerId>
        <ActionName>Pulshunt.Zuglaufen</ActionName>
        <Ziel>
          <KnotenId>Gleis8</KnotenId>
          <Kantenanfang>1</Kantenanfang>
          <Kantenende>2</Kantenende>
          <Abstand>61,218</Abstand>
        </Ziel>
      </Action>
      <Action>
      <Action>
      <Action>
      <Action>
    </Actions>
  </Scenario>
  <Fahrplan />
  <Guttypen>
</Projekt>
```

Figure 7-11: An example of the defined actions in the XML file in PULRAN

When the actions are defined, the potential conflicts with other trains are not considered. In the simulation process, the simulator will execute the actions that are read from the XML file. The conflicts and deadlocks will be automatically identified and resolved by the simulator. Although it is convenient for the users and developers of PULRAN to build complex shunting operations with the defined actions, the XML file used in PULRAN is not a real script language. It is not intuitive for normal users to learn the control structure of the XML files.

As a general-purpose scripting language, Python has been widely used for rapid and efficient development with a high level of readability. Based on the simulation tool PULSim, a prototype for the open interface of timetable simulation for Python developers is implemented (Figure 7-12). With this interface, Python developers can designate the investigated data of infrastructure, rolling stocks, and the timetable. The time period (in this example, from 7:00 to 8:00) for the timetable simulation can also optionally be specified. A timetable simulation will then be started with few lines of code.

```
01 from py4j.java_gateway import JavaGateway
02 gateway = JavaGateway()
03
04 print ("load data:")
05 # read infrastructure, rollingstocks and timetable data
06 simulator = gateway.entry_point.initialize(
07          "file:///c://temp//raildata//infrastructure",
08          "file:///c://temp//raildata//rollingstocks",
09          "file:///c://temp//raildata//timetable")
10
11 # the simulation period will be set from 7:00 to 8:00
12 gateway.entry_point.setSimulationTime(simulator, 7, 8)
13
14 simulator.run()
15
16 # the results of simulation can be customized by user
17 print ("Occupancy time of tracks:")
18 elements = gateway.entry_point.getAllInfrastructureElements()
19 for element in elements:
20     oMap = gateway.entry_point.getOccupancyMap(simulator, element)
21     if oMap is not None:
22         print ("Track Id: ", element, ":")
23         lastKey = 0;
24         for key in oMap:
25             if key != 0:
26                 print ("    ", lastKey, "-", key, "m:", oMap[key])
27             lastKey = key
28         print()
29
```

Figure 7-12: Execution of a timetable simulation in Python

After the timetable simulation has been carried out, the logged results of the simulation can be further read and evaluated. In Figure 7-12, the output of occupancy time for tracks is customised by the script written in Python. A part of the output of executing the Python code "TimetableSimulation.py" is shown in Figure 7-13. As described in Section 7.2.1, the occupancy time of a track is calculated separately for different segments, which are referenced by a starting position and an ending position.

```
Occupancy time of tracks:
Track Id:   82716.hh - 82714.hh :
      0.0 - 229.0 m: Duration [Seconds=1,172.600]

Track Id:   82521.hh - 84225.hh :
      0.0 - 356.0 m: Duration [Seconds=74.222]

Track Id:   84225.hh - 69940.hh :
      0.0 - 894.0 m: Duration [Seconds=707.227]

Track Id:   82468.hh - 82469.hh :

Track Id:   82466.hh - 83670.hh :
      0.0 - 721.0 m: Duration [Seconds=73.574]
    721.0 - 729.0 m: Duration [Seconds=202.148]

Track Id:   83062.hh - 83065.hh :
      0.0 - 153.0 m: Duration [Seconds=698.766]

Track Id:   82909.hh - 82819.hh :
      0.0 - 181.0 m: Duration [Seconds=64.659]
    181.0 - 211.0 m: Duration [Seconds=216.967]

Track Id:   83783.hh - 83434.hh :
      0.0 - 52.0 m: Duration [Seconds=216.967]

Track Id:   82875.hh - 82874.hh :

Track Id:   82877.hh - 82712.hh :
      0.0 - 322.0 m: Duration [Seconds=62.435]
    322.0 - 568.0 m: Duration [Seconds=98.536]
    568.0 - 618.0 m: Duration [Seconds=634.969]
```

Figure 7-13: Customised output of the calculated occupancy time for tracks

Supported by the open interface, a user-based adaptable simulation platform can be built. Furthermore, the script can be defined in different levels of detail, including the commands for an individual train run, a timetable simulation, and an operational simulation. For a sophisticated workflow, e.g., the process of calibrating disturbance parameters for operational simulation (see Section 6.4), the open interfaces can provide an open access to customise a third party algorithm flexibly. Users and researchers gain the ability to rapidly develop their own algorithm, which is not necessarily known by the simulation tool in advance. The productivity of using simulation tools for further evaluation and optimisation will be improved through the open interfaces significantly.

7.4　Graphical User Interface and Usability

The graphical user interface (GUI) of a railway simulation tool should provide an insightful view, and support an efficient user interaction. The design of GUI should focus on the visualisation and interaction concepts. In this section, the GUI and usability of the simulation software PULSim are addressed.

Concerning the visualisation concept, not only the movement of trains, but also the information and the interaction behind the scene of a simulation process should be presented. Therefore, the user interface is organised as different views in PULSim as shown in Figure 7-14.

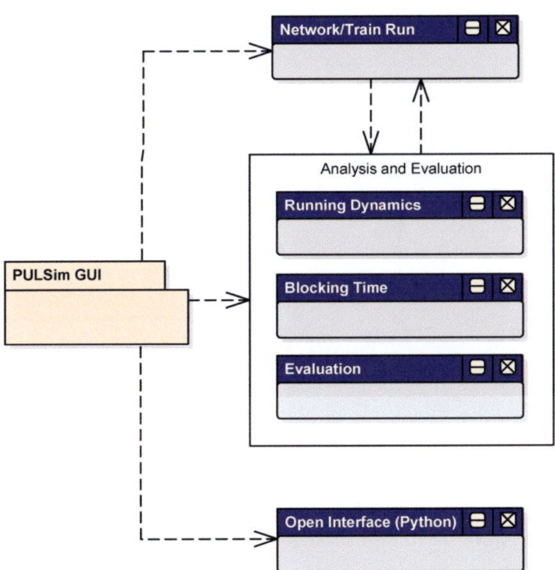

Figure 7-14: The views of GUI in PULSim

As a basic function of a simulation tool, the movement of train runs and the overview of simulated network should be presented to user, which is implemented in the "network/train run" view. In Figure 7-15, the screenshot of the "network/train run" view is illustrated. The control button can start, pause or stop a simulation. Trains are marked in green, red or yellow colour, which represents a train without hindrances, with hindrances due to occupancy conflicts, or with hindrances due to deadlocks, respectively. A navigator floating on top of the screen can be used to switch among different views.

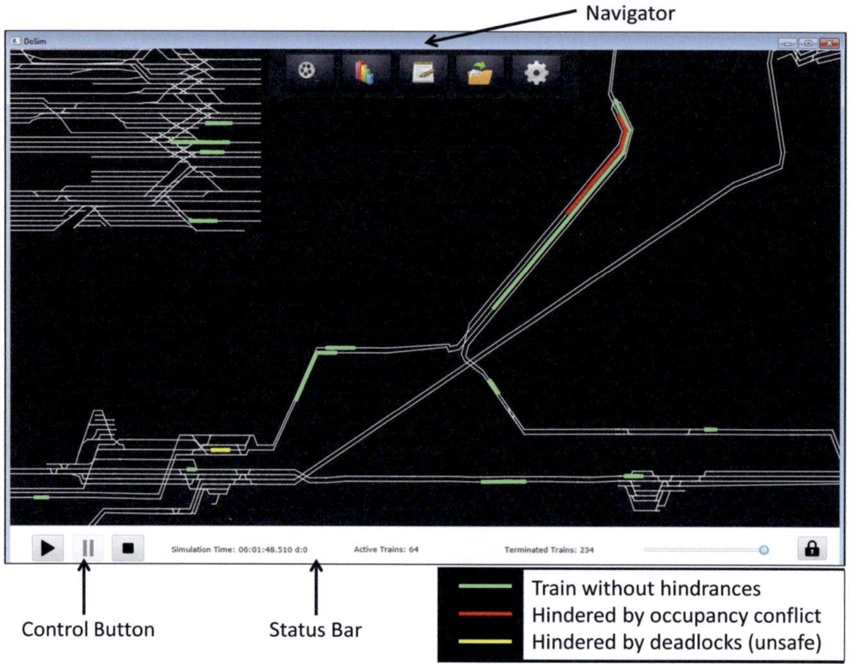

Figure 7-15: Screenshot of "network/train run" view in PULSim

At the current version, the views for analysis and evaluation are "running dynamics" view (see Figure 4-9), "blocking time" view (see Figure 4-11 and Figure 4-12), and "evaluation" view (see Figure 7-3 and Figure 7-4). The user can also trace the occurrence of events. In Figure 7-16, an example of tracing events for a train run is illustrated. User can click the interested point, at which one or several events take place. Meanwhile, the path of the train (marked as a blue line) and the exact location of the event (marked as a red point) is heighted in another window.

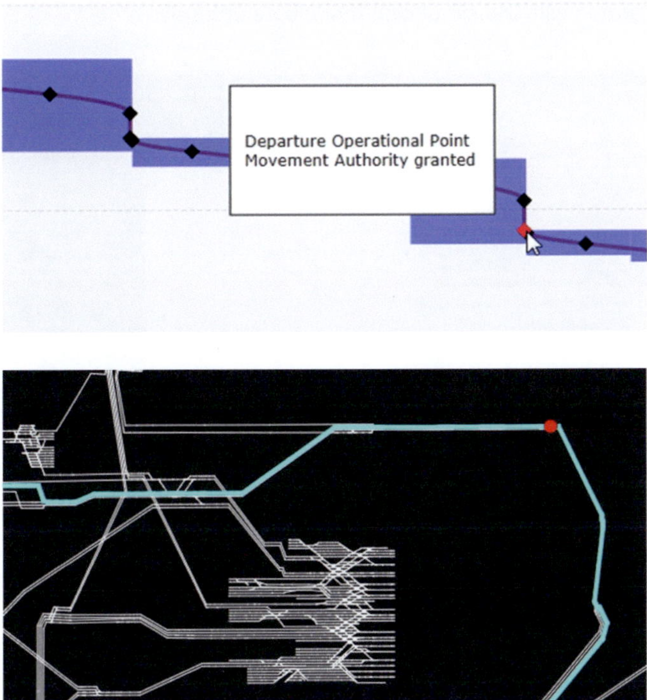

Figure 7-16: An example of tracing occurrence of events

Since a single round of simulation process is carried out automatically, the interventions from users during simulation are not required. Essentially, the interactions on analysis and evaluation of the generated simulation data are concerned. In the software PULSim, the detailed running dynamic of each train run as well as the events and conflicts encountered during a simulation process are recorded, which provide the analyse and evaluation modules with a comprehensive data basis with a high performance. From the views for analysis and evaluation, the information can be presented to users from different perspectives. Users can switch from the different views flexibly.

Furthermore, the open interface for Python has been integrated into PULSim (see Figure 7-17). The same code shown in Section 7.3.3 can be input, edited, saved and executed in the integrated development environment. In Figure 7-17, the Python code is input in the left area (the editor), and the result is output in the right area (the information window). Hence, a user-based analysis and evaluation can be customised and integrated with the simulation module seamlessly.

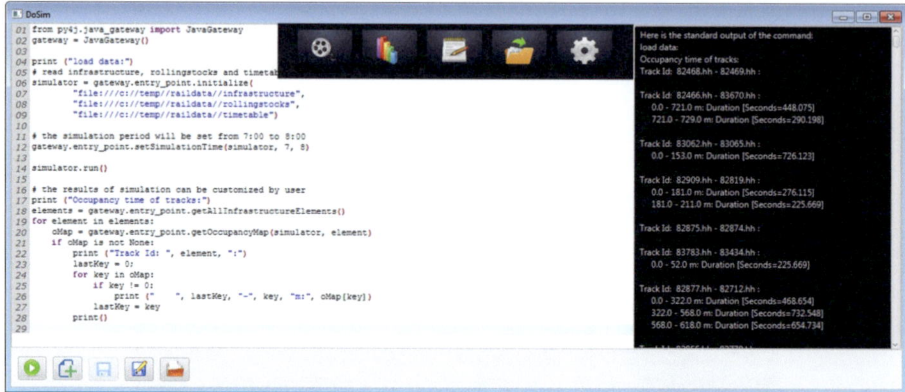

Figure 7-17: Screenshot of the GUI for open interfaces with Python in PULSim

8 Conclusion and Further Research and Development

Simulation approaches can provide an efficient way to evaluate different design alternatives for railway planning and operations in an inexpensive way. The results of simulation comprise the basis data for further evaluation and optimisation. A simulation model can be perceived as a representation of the actual system. In addition, the facts that are unseen in reality can be revealed by simulation approaches through analysing the output of a simulation.

Today, several simulation tools are available for railway planning and operations. In this book, the design details of the simulation tool PULSim are presented, in order to provide readers with a deep understanding of the structural and behavioural aspects of railway simulation design and analysis. Before a project with simulation approaches is undertaken, the knowledge of the simulation model is always the most important prerequisite.

Research and development in the field of railway simulation and analysis is still very active. For further research and development, the validation and calibration methods, the use cases of simulation approaches, as well as the technology for simulation analysis should be stressed for railway planning and operations.

As discussed in Chapter 6, the credibility and the correctness of a simulation model are the necessary conditions for success. The methods for automatic calibration and validation are indispensable for simulation applications. Today, more and more data are available from different data sources, which are generated and collected during daily railway operations. These data can be used to validate the credibility of simulation tools. The innovative technologies of big data are helpful to capture, store, query, and analyse the data, so that the quality and accuracy of the simulation model can be improved. Through learning from the actual data, the applied simulation model can simulate design alternatives that are close to reality.

Today, most applications of railway simulation focus on the evaluation of simulation results. The simulation approach is often utilised in the form of experiments. It is a trend to shift the applications of railway simulation from evaluation to the fields of simulation-based optimisation. During iterative processes of changing the design, solutions for railway planning and operations advance towards their optimal values. In traditional design and planning processes, the design alternatives are derived by experienced users and experts. With simulation-based optimisation, the optimum solutions for infrastructure planning, arrangement of signalling systems and scheduling can be found automatically. Similarly, during railway operations, the control of driving style and the dispatching decisions can be optimised with the support of the simulation approach.

For a complex system in a large scale network, it is common to simulate railway planning and operations in several different packages, including timetabling, rolling stock assignment, and crew scheduling. The processes for planning and the decision making for dispatching rolling stocks and crews are also carried out separately. A unified simulation platform can integrate these processes seamlessly based on a same model. Furthermore, the framework of multi-scale simulation and distributed computing architecture will facilitate research for large-scale networks with a high level of accuracy.

In recent years, artificial intelligence and machine learning have been successfully applied in many branches of study. The hidden logic and the philosophy for railway planning and operations can be investigated and learned. Thanks to the increased level of automation, including automatic train protection and automatic train operation, the system behaviour and the impacts of design can be predicted more exactly. During the planning and decision-making processes, the simulation approach can support data mining and deep learning processes either online or offline. It possesses great potential to combine cutting-edge technology with railway simulation to improve the quality and efficiency of railway planning and operations.

Publication bibliography

Albrecht, T. (2014): Energy-efficient railway operation. In I. A. Hansen, Jörn Pachl (Eds.): Railway Timetabling & Operations. Analysis Modelling Optimisation Simulation Performance Evaluation. Hamburg: Eurailpress DW Media Group, pp. 91–116.

Anders, E.; Theeg, G.; Vlasenko, S. V. (2009): Railway Signalling & Interlocking. International Compendium. 1st ed. Hamburg: Eurailpress.

Bangsow, S. (2010): Manufacturing Simulation with Plant Simulation and SimTalk. Berlin, Heidelberg: Springer Berlin Heidelberg.

Brünger, O.; Dahlhaus, E. (2014): Running time estimation. In I. A. Hansen, Jörn Pachl (Eds.): Railway Timetabling & Operations. Analysis Modelling Optimisation Simulation Performance Evaluation. Hamburg: Eurailpress DW Media Group, pp. 65–89.

Büker, T.; Seybold, B. (2012): Stochastic modelling of delay propagation in large networks. In *Journal of Rail Transport Planning & Management* 2 (1-2), pp. 34–50.

Bussieck, M. R.; Winter, T.; Zimmermann, U. T. (1997): Discrete optimization in public rail transport. In *Mathematical Programming* 79 (1-3), pp. 415–444. DOI: 10.1007/BF02614327.

Caprara, A.; Fischetti, M.; Toth, P. (2002): Modeling and solving the train timetabling problem. In *Operations Research* 50 (5), pp. 851–861. DOI: 10.1287/opre.50.5.851.362.

Coffman, E. G.; Elphick, M.; Shoshani, A. (1971): System deadlocks. In *ACM Computing Surveys* 3 (2), pp. 67–78. DOI: 10.1145/356586.356588.

Cui, Y. (2009): Spezification of Project PULRAN - Programm zur Untersuchung der Logistik im Rangierdienst.

Cui, Y. (2010): Simulation-Based Hybrid Model for a Partially-Automatic Dispatching of Railway Operation. (Simulationsbasiertes Hybrid-Modell für eine Teilautomatisierte Disposition des Eisenbahnbetriebes). Norderstedt: Books on Demand GmbH (Neues verkehrswissenschaftliches Journal, 4).

Cui, Y.; Krohn, T., Martin, U.; Tritschler, S. (2012): Consistent decision process and algorithm for train dispatching. In V. Schwieger, F. Englmann, M. Friedrich, J. Jessen,

R. Large, U. Martin et al. (Eds.): 6th International Symposium, Proceedings. Networks for Mobility. Stuttgart, Germany, pp. 59–60.

Cui, Y.; Martin, U. (2011): Multi-scale simulation in railway planning and operation. In *PROMET* 23 (6). DOI: 10.7307/ptt.v23i6.186.

Cui, Y.; Martin, U.; Liang, J. (2017): Decentralised, Autonomous Train Dispatching using Swarm Intelligence in Railway Operations and Control. In : 7th International Conference on Railway Operations Modelling and Analysis. Lille.

Cui, Y.; Martin, U.; Zhao, W. (2016): Calibration of disturbance parameters in railway operational simulation based on reinforcement learning. In *Journal of Rail Transport Planning & Management* 6 (1), pp. 1–12. DOI: 10.1016/j.jrtpm.2016.03.001.

Cui, Y.; Martin, U.; Zhao, W.; Cao, N. (2014): Entwicklung eines Algorithmus für die Kalibrierung von Modellen zur Betriebssimulation in Spurgeführten Verkehrssystemen unter Berücksichtigung Stochastischer Bedingungen. DFG-Forschungsprojekt (MA 2326/9-1). Norderstedt: Books on Demand (Neues verkehrswissenschaftliches Journal, 9).

Daamen, W.; Goverde, R. M. P.; Hansen, I. A. (2008): Non-discriminatory automatic registration of knock-on train delays. In *Networks and Spatial Economics* 9 (1), pp. 47–61.

DB Netz AG (2008): Richtlinie 405. Fahrwegkapazität.

Dijkstra, E. W. (1982): The mathematics behind the Banker's algorithm, Selected Writings on Computing: A Personal Perspective. In Edsger W. Dijkstra (Ed.): Selected Writings on Computing: A Personal Perspective. New York: Springer-Verlag, pp. 308–312.

Draper, N. R.; Smith, H. (1998): Applied Regression Analysis. 3rd ed. New York, Chichester: Wiley (Wiley series in probability and statistics).

Ernst, A. T.; Jiang, H.; Krishnamoorthy, M.; Nott, H.; Sier, D. (2001): An integrated optimization model for train crew management. In *Annals of Operations Research* 108, pp. 211–224. DOI: 10.1023/A:1016019314196.

Fabris, S. de; Longo, G.; Medeossi, G. (2008): Automated analysis of train event recorder data to improve micro-simulation models. In J. Allan, R. J. Hill, C. A. Brebbia,

G. Sciutto, S. Sone (Eds.): Computers in Railways XI. Southampton: WIT Press, pp. 575–583.

Fioole, P.-J.; Kroon, L.; Maróti, G.; Schrijver, A. (2006): A rolling stock circulation model for combining and splitting of passenger trains. In *European Journal of Operational Research* 174 (2), pp. 1281–1297. DOI: 10.1016/j.ejor.2005.03.032.

Gamma, E. (1995): Design Patterns. Elements of reusable object-oriented software. Wokingham: Addison-Wesley (Addison-Wesley professional computing series).

Ghoseiri, K.; Szidarovszky, F.; Asgharpour, M. J. (2004): A multi-objective train scheduling model and solution. In *Transportation Research Part B: Methodological* 38 (10), pp. 927–952. DOI: 10.1016/j.trb.2004.02.004.

Giaccone, M.; Pomposiello, P. (2011): AIMESC Project Final Report and Guidelines. ERTMS - a new technology for the railway sector, Anticipating its IMpact on Employment and Social Conditions. Brussels: European Transport Workers' Federation.

Goverde, R. M. P. (2005): Punctuality of railway operations and timetable stability analysis. Delft University of Technology Delft, Delft.

Goverde, R. M. P.; Daamen, W.; Hansen, I. A. (2008): Automatic identification of route conflict occurances and their consequences. In J. Allan, R. J. Hill, C. A. Brebbia, G. Sciutto, S. Sone (Eds.): Computers in Railways XI. Southampton: WIT Press, pp. 473–482.

Hansen, I. A.; Pachl, J. (2014): Railway Timetabling & Operations. Analysis Modelling Optimisation Simulation Performance Evaluation. Hamburg: Eurailpress DW Media Group.

Horn, P.; Zinke, R.: Energy-optimal control of trains in long distance traffic. In : Technical University of Budapest 1990 – Proceedings of the 2nd Mini Conference on Vehicle System Dynamics, pp. 170–184.

Howlett, P. G.; Pudney, P. J. (1995): Energy-Efficient Train Control. London: Springer London (Advances in Industrial Control).

Hürlimann, D. (2010): Open Track Betriebssimulation von Eisenbahnnetzen Version 1.6. Zürich.

IEEE (2004): Communications Based Train Control (CBTC). New York: IEEE Rail Transit Vehicle Interface Standards Committee (Std 1474).

Kaner, C.; Falk, J. L.; Nguyen, H. Q. (1999): Testing Computer Software. 2nd ed. New York, Chichester: John Wiley (Wiley computer publishing).

Kecman, P.; Goverde, R. M. P. (2012): Process mining of train describer event data and automatic conflict identification. In C. A. Brebbia, N. Tomii, J. M. Mera (Eds.): Computers in Railways XIII, WIT Transactions on the Built Environment. Southampton: WIT Press, pp. 227–238.

Kecman, P.; Goverde, R. M. P. (2015): Predictive modelling of running and dwell times in railway traffic. In *Public Transport* 7 (3), pp. 295–319.

Kogel, B.; Nießen, N.; Büker, T. (2010): Influence of the European Train Control System (ETCS) on the Capacity of Nodes. Paris.

Kroon, L.; Huisman, D.; Maroti, G. (2014): Optimisation models for railway timetabling. In I. A. Hansen, Jörn Pachl (Eds.): Railway Timetabling & Operations. Analysis Modelling Optimisation Simulation Performance Evaluation. Hamburg: Eurailpress DW Media Group, pp. 155–190.

Law, A. M. (2007): Simulation Modeling and Analysis. 4th ed. Boston, London: McGraw-Hill (McGraw-Hill series in industrial engineering and management science).

Lehmer, D. H. (1951): Mathematical methods in large-scale computing unit. In *The Annals of the Computation Laboratory of Harvard University* 26, pp. 141–146.

Liang, J. (2017): Metaheuristic-based Dispatching Optimization Integrated in Multiscale Simulation Model of Railway Operation. Dissertation, Stuttgart. Institut für Eisenbahn- und Verkehrswesen der Universität Stuttgart.

Ligier, D.; Guido, P. (2012): ERTMS Operational Principles and Rules. Version 2.1.8. Valenciennes: European Railway Agency (ERTMS UNIT).

Liu, R.; Golovitcher, I. M. (2003): Energy-efficient operation of rail vehicles. In *Transportation Research Part A: Policy and Practice* 37 (10), pp. 917–932. DOI: 10.1016/j.tra.2003.07.001.

Luethi, M.; Huerlimann, D.; Nash, A. (2005): Understanding the timetable planning process as a closed control loop. In I. A. Hansen, F. M. Dekking, R. M. P. Goverde, B.

Heidergott, L. E. Meester (Eds.): Proceedings of the 1st International Seminar on Railway Operations Modelling and Analysis. Delft.

Markovića, N.; Milinkovićb, S.; Tikhonovc, K. S.; Schonfelda, P. (2015): Analyzing passenger train arrival delays with support vector regression. In *Transportation Research Part C: Emerging Technologies* 56, pp. 251–262.

Martin, U. (1995): Erfahren zur Bewertung von Zug- und Rangierfahrten bei der Disposition. Dissertation. Technischen Universität Braunschweig.

Martin, U. (2015): Railway operation, multimodal transport planning & modelling. Lecture Notes. Infrastructure Planning (M.Sc.). Universität Stuttgart, 2015.

Martin, U. (2016): Durchführung des Betriebes, Grundlagen der Verkehrssysteme. Lecture Notes. Produktion und Absatz von Verkehrsleistungen. Universität Stuttgart, 2016.

Martin, U.; Cui, Y.; Hantsch, F.; Chu, Z.; Li, X. (2014): Knotenkapazität - Bewertungsverfahren für das Mikroskopische Leistungsverhalten und die Engpasserkennung im Spurgeführten Verkehr (RePlan). Norderstedt: Books on Demand (Neues verkehrswissenschaftliches Journal, 8).

Martin, U.; Schmidt, C.; Chu, Z. (2011): PULEIV Anwendungsleitfaden zur PULEIV-Version 2.1. Anleitung zur Ermittlung des Leistungsverhaltens von Eisenbahninfrastruktur mithilfe von Simulationsprogrammen.

Medeossi, G.; Longo, G.; Fabris, S. de (2011): A method for using stochastic blocking times to improve timetable planning. In *Journal of Rail Transport Planning & Management* 1 (1), pp. 1–13.

Melle, A.; Lübbe, C.; Pfitzner, K.; Wörner, M.; Weissinger, T. (2006): Spezifikation - Studienprojekt Marian 2. IEV, Universität Stuttgart. Stuttgart.

Mills, G.; Pudney, P. (2008): The effects of deadlock avoidance on rail network capacity and performance. In : Proceedings of the 2003 Mathematics-in-Industry Study Group. Mathematics-in-Industry Study Group. Australian Mathematical Society, pp. 49–63.

Mittermayr, R.; Blieberger, J.; Schöbel, A. (2012): Kronecker algebra-based deadlock analysis for railway systems. In *PROMET* 24 (5), pp. 359–369. DOI: 10.7307/ptt.v24i5.1171.

Neuber, M. (2017): Ein Beitrag zur operativen Zuglaufregelung unter besonderer Berücksichtigung vorausschauender Zugförderung bei der Umsetzung dispositiver Entscheidung. Dissertation. Institut für Eisenbahn- und Verkehrswesen der Universität Stuttgart.

Pachl, J. (1993): Steuerlogik für Zuglenkanlagen zum Einsatz unter stochastischen Betriebsbedingungen. Dissertation, Braunschweig. Institut für Verkehr, Eisenbahnwesen und Verkehrssicherung, Technische Universität Braunschweig.

Pachl, J. (2007): Avoiding deadlocks in synchronous railway simulations. In : 2nd International Seminar on Railway Operations Modelling and Analysis. Proceeding CD-ROM. Hannover, Germany, March 28 - 30.

Pachl, J. (2011): Deadlock avoidance in railroad operations simulations. Paper No. 11-0175. In Transportation Research Board (Ed.): TRB 90th Annual Meeting Compendium of Papers DVD. Washington, DC.

Petersen, E. R.; Taylor, A. J. (1983): Line block prevention in rail line dispatch and simulation models. In *Information Systems and Operations Research* 21 (1), pp. 46–51.

Pudney, P.; Wardrop, A. (2008): Generating train plans with problem space search. In Mark Hickman, Pitu Mirchandani, Stefan Voß (Eds.): Computer-aided Systems in Public Transport, vol. 600. Berlin, Heidelberg: Springer Berlin Heidelberg (Lecture Notes in Economics and Mathematical Systems), pp. 195–207.

Radtke, A. (2005): EDV-Verfahren zur Modellierung des Eisenbahnbetriebs. Hamburg: Eurailpress Tetzlaff-Hestra (Wissenschaftliche Arbeiten für den Schienenverkehr, 64).

Radtke, A. (2014): Infrastructure modelling. In I. A. Hansen, Jörn Pachl (Eds.): Railway Timetabling & Operations. Analysis Modelling Optimisation Simulation Performance Evaluation. Hamburg: Eurailpress DW Media Group, pp. 47–63.

Radtke, A.; Bendfeldt, J. (2001): Handling of railway operation problems with RailSys. In : Proceedings of the 5th World Congress on Rail Research. Cologne, Germany.

railML (2013): IS:TrackConnections - Wiki.railML.org. Available online at http://wiki.railml.org/index.php?title=IS:TrackConnections, updated on 2/10/2013, checked on 12/3/2015.

Riedmiller, M.; Braun, H. (1993): A direct adaptive method for faster backpropagation learning: The RPROP algorithm. In : Proceedings of IEEE International Conference on Neural Networks. San Francisco.

RMCON (2010): RailSys 7 User Manual. Preparations for Multiple Simulation.

Schmidt, C. (2009): Beitrag zur Experimentellen Bestimmung der Wartezeitfunktion bei Leistungsuntersuchungen im Spurgeführten Verkehr. Norderstedt: Books on Demand GmbH (Neues verkehrswissenschaftliches Journal, 3).

Schwanhäußer, W. (1978): Die Ermittlung der Leistungsfähigkeit von großen Fahrstraßenknoten und von Teilen des Eisenbahnnetzes. In *Archiv für Eisenbahntechnik : AET* 33 (1), pp. 7–18.

Serafini, P.; Ukovich, W. (1989): A mathematical model for periodic scheduling problems. In *SIAM Journal on Discrete Mathematics* 2 (4), pp. 550–581. DOI: 10.1137/0402049.

Seybold, B.; Huerlimann, D. (2016): OpenTrack - Simulation of railway systems. Presentation of the OpenTrack API. OpenTrack. Available online at http://www.opentrack.ch/opentrack/downloads/OpenTrack.API.pdf, checked on 8/1/2016.

Sipilä, H. (2015): Simulation of Rail Traffic. Methods for Timetable Construction, Delay Modeling and Infrastructure Evaluation. Stockholm: Architecture and the Built Environment (TRITA-TSC-PHD, 15:001).

Strobel, H.; Horn, P.; Kosemund, M. (1974): Contribution to optimum computer - aided control of train operation. In : 2nd IFAC/IFIP/IFORS World Symposium "Traffic Control and Transportation Systems". Monte Carlo, pp. 377–387.

Tsichritzis, D.; Klug, A. (1978): The ANSI/X3/SPARC DBMS framework report of the study group on database management systems. In *Information Systems* 3 (3), pp. 173–191. DOI: 10.1016/0306-4379(78)90001-7.

UIC (2013): Capacity. 2nd Edition. Paris: International Union of Railways (UIC code, 406).

Vakhtel, S. (2002): Rechnerunterstützte Analytische Ermittlung der Kapazität von Eisenbahnnetzen. Aachen: Verkehrswissenschaftl. Institut RWTH Aachen (Veröffentlichungen des Verkehrswissenschaftlichen Instituts der Rheinisch-Westfälischen Technischen Hochschule Aachen, 59).

VDV (2013): VDV-Schrift 452: VDV-Standardschnittstelle Liniennetz/Fahrplan. ÖPNV Datenmodell 5.0-„Schnittstellen-Initiative". Available online at https://www.vdv.de/oepnv-datenmodell.aspx, checked on 12/7/2015.

VIA Con (2011): LUKS Handbuch Version 2.1.

Voss, S.; Daduna, J. R. (2001): Computer-aided scheduling of public transport. Berlin, New York: Springer (Lecture Notes in Economics and Mathematical Systems, 505).

Waston, R.; Medeossi, G. (2014): Simulation. In I. A. Hansen, Jörn Pachl (Eds.): Railway Timetabling & Operations. Analysis Modelling Optimisation Simulation Performance Evaluation. Hamburg: Eurailpress DW Media Group, pp. 191–215.

Wörner, M.; Cui, Y. (2008): Begriffslexikon - Datenmodell. Institut für Eisenbahn- und Verkehrswesen der Universität Stuttgart. Stuttgart.

Yuan, J. (2006): Stochastic modelling of train delays and delay propagation in stations. Ph.D. Thesis. Delft University of Technology Delft, Delft.

Zhao, W.; Martin, U.; Cui, Y.; Kösters, M. (2016): Calibration of timetable parameters for rail-guided systems. In *IJR - International Journal of Railway* 9 (1), pp. 1–9.

Glossary

Asynchronous simulation

In asynchronous simulation, trains are introduced into the simulation system according to their pre-defined priorities, respectively. The train introduced firstly has a higher priority and should not be influenced by lower priority trains. The lower priority trains are inserted to the system after the train runs of higher priority trains have been fixed.

Authorised running section

During a simulation process, a round of calculation of running dynamics is carried out for the section from the actual position of the head of the train to the EOA. This section is defined as an authorised running section.

Basic structure

A basic structure is defined as a basic occupancy element in which all parts should be equally occupied, regardless of the design of the operating program. Therefore, it is prohibited for two or more trains to run simultaneously upon a basic structure. Basic structures can be categorised as junction-type and non-junction-type. A junction-type basic structure contains at least one junction-type infrastructure element, while a non-junction-type basic structure only contains a track or a part of track.

Behaviour section

Behaviour section is a general term used to describe the behaviour (acceleration, cruising, coasting, and braking) at one characteristic section.

Calibration

The process of validating and correcting the parameters used in a simulation model is called calibration.

Characteristic section

Within a moving section, one or several characteristic sections are defined. A characteristic section describes a part of infrastructure with identical geometrical attributes and maximum allowed velocity. Hence the calculation of running dynamics for a characteristic section can be handled with at most one full operational cycle of behaviour sections.

Complex infrastruc-

A complex infrastructure object serves as the combination of

ture object	the above point-shaped objects and line-shaped objects. A complex infrastructure object contains at least two line-shaped objects, with or without point-shaped objects.
Deadlock	A deadlock is a self-blockade in a control system in which two or more tasks are waiting for each other to release a resource in a circular chain. In rail traffic, a deadlock is a situation in which a number of trains cannot continue their path at all because every train is blocked by another one.
Definition of train formation	A train can consist of locomotives, wagons or multiple units. After the formation of a considered train is defined in the form of a definition of train formation, the attributes of said train can be aggregated from the included vehicle definitions.
Discrete point	A series of discrete points are used to describe the running dynamics of the entire train run. Each discrete point contains the following information:

- Distance from the last discrete point to the current point
- Running time from the last discrete point to the current point
- Velocity at the current point

Disturbance parameter	For an operational simulation, it is decisive to generate realistic disturbances as close as possible to the reality. The disturbances are modelled through a set of statistical parameters, which are called disturbance parameters
End of authority	The End Of Authority (EOA) is the location to which the train is authorised to process and where the target speed is zero
Entry	In a timetable, the arrival/departure/passing time at each station is defined and is stored in an entry. A timetable consists of a series of entries sorted by the scheduled arrival/departure/passing time.
Event	An event is defined as an occurrence that may change the state of the system at a certain time point. Although the attributes of the involved components (infrastructure, rolling stocks, and operations) may be changed continuously during a simu-

	lation process, only the specific occurrence of the changing of state, which causes interactions between two or more components or triggers another event, is considered a discrete event.
Event-driven simulation	In event-driven simulation, the system evolves over a series of time points. At minimum, one certain event occurs at each time point. While the time interval between two time points is fixed in time-driven simulation, the events can be scheduled realistically according to the exact occurrence time in event-driven simulation.
False-positive	In a "false-positive" situation, a request, which may not lead to a deadlock, will be rejected as the result of the deadlock-free test failing to indicate a safe state.
Feasible resource	Feasible resources are one or more connected infrastructure resources in the further path of the tested train. If feasible resources are available, the infrastructure from the requested resources to the feasible resources can be granted to the tested train, even the train has not passed the deadlock-free test.
Immediate conflict	The immediate conflict is defined as an occupancy conflict that takes place at a position at which there is no other occupancy conflict for both conflicting trains. The nearest occupancy conflict between two train runs, which is known as an immediate conflict, is the highest priority concern for a dispatching system.
Indicator	In a calibration process, the term "indicator" refers to the data collected in reality. The simulated results will be compared with the indicators in order to verify them and to calibrate the parameters used in a simulation model.
Infrastructure element model	The infrastructure elements, including tracks, turnouts and crossings etc., are used as the fundamental building blocks in the model. An infrastructure element is a complex infrastructure object. Inside an infrastructure element, each endpoint is

	modelled as a port, and two ports are connected via several directed edges, which indicate the possibilities of train runs.
Infrastructure object	An infrastructure object is a generic term for all the objects used in infrastructure modelling. It can represent a real object (e.g. signals, tracks or turnouts) or an abstract object (e.g. paths, block sections)
infrastructure resource	An infrastructure resource is defined as the minimum set of infrastructure in railway operations, which is requested, allocated, and released as a basic unit at the same time.
Line-shaped infrastructure object	A line-shaped infrastructure object is connected by two or more point-shaped objects in a certain sequence. It can only consist of either two mono-directional objects or two bidirectional objects, but not a combination of the types.
Macroscopic level	On a macroscopic level, the components of a railway system are modelled at the highest abstraction level for a large scale of the investigated area with a very large number of train movements.
Mesoscopic level	Depending on the application context, a mesoscopic level can be applied for a model that integrates both the macroscopic and microscopic levels, or for a model between macroscopic level and microscopic level with respect to its level of detail.
Microscopic level	A model on a microscopic level describes the most detailed information with high accuracy. It can be used to simulate railway operations with the highest level of detail.
Movement authority	A movement authority is the permission given to a train to move up to defined position.
Moving section	A moving section represents an independent part for the calculation of running dynamics inside of an authorised running section.
Open track section	The term "open track section" can be used to specify a link at the macroscopic level. An open track section provides connections between two nodes. In case one side of the open track section terminates in a dead-end of the network, it will

	then connect only to one node. Depending on the signalling system and operation conditions, an open track section can be mono-directional or bidirectional. Inside a node, the possible directions of paths can be determined based on the directions of the connected open track sections.
Operating program	An operating program is comprised of all the details of railway operations, which justify the construction of railway infrastructure and determine infrastructure layout.
Parts of train run	Sometimes a passenger train can be decoupled into two sub-trains flexibly with coupling and sharing operations. In this case, several Parts of Train Run (PTR) should be defined under a train run. Each part of train run is operated along a specific path with a sub-train number, and the definition of train formation for each part of train run should be specified.
Path component	A path component is a directed edge inside a block section used for train runs, which can be released separately as a directed occupancy element. A block section consists of one or several path components (if partial route releasing is applied). An overlap will be modelled as a path component for the signalling system with overlaps
Pending time	The pending time is defined as the time period from the time at which the requested infrastructure is supposed to be allocated to the train in case of no hindrances, to the actual time that the train obtains the movement authority to enter the requested infrastructure resource.
Point-shaped infrastructure object	A point-shaped infrastructure object models the concerned object as a point. Some examples of point-shaped infrastructure objects include signals, route releasing points, or stations.
Reference point	The term "reference point" of a timetable is defined as a general term for a stop or a measurement point, and each entry is referenced by a reference point.
Roster of a train	A roster of a train models the usage of the train during a day-to-day operation, and begins and ends at a depot. A roster of

	a train consists of several sub-rosters of the train. For each sub-roster of the train, a part of train run is assigned, which is bound to the same definition of train formation of the concerned train.
Roster of a crew	For each staff member, a crew roster is created, which includes several sub-rosters of the crew. One or more sub-rosters of the crew are assigned for a sub-roster of a train. A sub-roster of crews can start or end at a depot or a stop.
Station tracks	Station tracks connect one or more junction-type basic structures inside a station. With junction-type basic structures and station tracks, a station or a junction can be represented in the form of a node-link model at a mesoscopic level, wherein the nodes of a mesoscopic model are junction-type basic structures, and the links are station tracks.
Synchronisation Times	Synchronisation Times serve to the traffic-related adjustment of several trains to establish the connections and to adjust the departure time of each train in regular intervals. Synchronisation times are provided generally as an extension of the dwell time. It is possible to allocate the synchronisation times to the running times. However, it will lead to a running time extension.
Synchronous simulation	Synchronous simulation is a dynamic simulation, in which all train runs are simulated simultaneously along the passage of time.
Time-driven simulation	In time-driven simulation, all of the train runs and the interaction between trains and infrastructure are simulated synchronously along a series of time points. The time interval between two given time points is fixed.
Train run	A timetable defines a number of train runs, whereby each train run has a fixed train number as an identity during operations.
Unit definition	The general term "unit definition" is used as the super class of vehicle definition and definition of train formation. The unit definition only provides the basic attributes shared by vehicle

	definitions and definitions of train formation.
Validation	The term "validation" refers to the process of comparing the simulated results and the data collection in reality.
Vehicle definition	The attributes of a vehicle definition consist of both general and specific attributes. General attributes describe the information required for all application contexts. A set of specific attributes describes a certain configuration of the vehicle definition.
Verification	The term "verification" refers to the process of determining whether a model has been implemented correctly in computer software. The simulation results will be compared with the expected value, which is usually derived from a theoretical calculation.
Waiting time	Waiting time is defined as the extension of running time and dwell time due to hindrances from other trains (DB Netz AG 2008). Waiting time can be categorised as unscheduled waiting time and scheduled waiting time. Unscheduled waiting time is known as the delay, which is caused by the stochastic disturbances during railway operations. Scheduled waiting time is the artificial increase in the overall time of a train caused by the resolution of conflicts during the scheduling process (Hansen, Pachl 2014).

Abbreviations

API	Application Programming Interface
ATP	Automatic Train Protection
AR	Currently Available Resources
AWS	Automatic Waring System
CBTC	Communication-Based Train Control
CSV	Comma-Separated Values
DRR	Dynamic Route Reservation
EOA	End OF Authority
ETCS	European Train Control System
FCFS	First-Come First-Served
FCLS	First-Come Last-Served
HTTP	Hypertext Transfer Protocol
ICE	Intercity Express
LCG	Linear Congruential Generators
IEV	Institute for Railway and Transportation Engineering, in German: Institut für Eisenbahn- und Verkehrswesen der Universität Stuttgart
LTS	Least Trimmed Squares
LZB	Continuous Train Protection, in German: Linienzugbeeinflussung
MCA	Movement Consequence Analysis
MR	Maximal Required Resources
OCC	Operations Control Centre
OR	Currently Occupied Resources
ORM	Object-Relational Mapping
P	A set of trains, which contains all the trains that are satisfied with the needs of their requested resources
PESP	Periodic Event Scheduling Problem
PTR	Parts of Train Run
PULEIV	Program to Capacity Research, in German: Programm zur Untersuchung des Leistungsverhaltens

PULRAN   Program to Research the Logistic in the Shunting Operation, in German: Programm zur Untersuchung der Logistik im Rangierdienst

PZB   Intermittent Train Protection in German: Punktförmige Zugbeeinflussung

RPROP   Resilient Backpropagation

SOAP   Simple Object Access Protocol

SQL   Structured Query Language

TPWS   Train Protection and Warning System

UML   Unified Modelling Language

VDV   Association of German Transport Companies, in German: Verband Deutscher Verkehrsunternehmen

XML   Extensible Markup Language

Appendix A: Establish Junction-type Basis Structures and Station Tracks for Mesoscopic Modelling

The presented process of the establishment of junction-type basic structures and station tracks is based on the microscopic model as discussed in Section 2.2.4.

As first, the junction-type infrastructure elements (turnouts, crossings, single slips and double slips) are processed, in order to establish the set of junction-type basic structures. Then the station tracks based on the established junction-type basic structures should be built up. After the establishment of junction-type basic structures and station tracks, the infrastructure network can be modelled at a mesoscopic level.

A junction-type basic structure contains all the connected infrastructure elements, which can be junction-type or non-junction-type. If two connected infrastructure elements are not allowed to be occupied by two trains simultaneously, these infrastructure elements should belong to a same junction-type basic structure. The procedure of establishing junction-type basic structures is indeed to assign infrastructure elements to junction-type basic structures. It will be carried out until each junction-type infrastructure element has been assigned to a junction-type basic structure. The following algorithm is developed to establish junction-type basic structures:

1) A junction-type infrastructure element J that has not been assigned to a junction-type basic structure will be chosen. A junction-type basic structure R is initialised, which contains the infrastructure element J.

2) A set S that includes all the neighbouring infrastructure elements of the junction-type basic structure R should be created.

3) Each neighbouring infrastructure element N inside the set S will be checked. If all neighbouring infrastructure elements are examined, proceed to Step 7.

4) If the neighbouring infrastructure element N is junction-type, proceed to Step 5. If N is a non-junction-type infrastructure element (a track), the signalling system of the investigated network should be taken into consideration. For a non-signalised network or a moving block system, the infrastructure element N will not be processed, return to Step 3. For a signalised network, if there is no signal or releasing point along the infrastructure element N, N will be added to the junction-type basic structure R. Return to Step 2.

5) A fictive junction-type basic structure R' will be created, which includes all the infrastructure elements in R and the current checked junction-type neighbour N. All the external ports, which are located on the boundary of the fictive junction-type basic structure R', will be identified. All the connections between two external ports should be created.

6) If there are two connections without overlapping, the junction-type element N is not allowed to be assigned to R. If all the connections overlap with each other, N will be added to R. Return to Step 2.

7) If all of the junction-type infrastructure elements have been assigned to a junction-type basic structure, the process should be terminated. Otherwise, return to Step 1.

The key idea behind the algorithm is to examine the overlapping situation. In Figure A 1, the overlapping situation of a fictive junction-type basic structure R' including turnout A and B is examined. The turnout A and B are connected through the port A1 and B1. Here a port is named according to the name of the turnout and the number of the port. For example, A1 represents port 1 of turnout A. The external ports for the checked junction-type basic structure R' are A2, A3, B2, and B3. All the connections of the external ports are A3↔B3, A2↔B3, A2↔B3, and A3↔B2. Since all the connections overlap each other, turnouts A and B can be assigned to a junction-type basic structure.

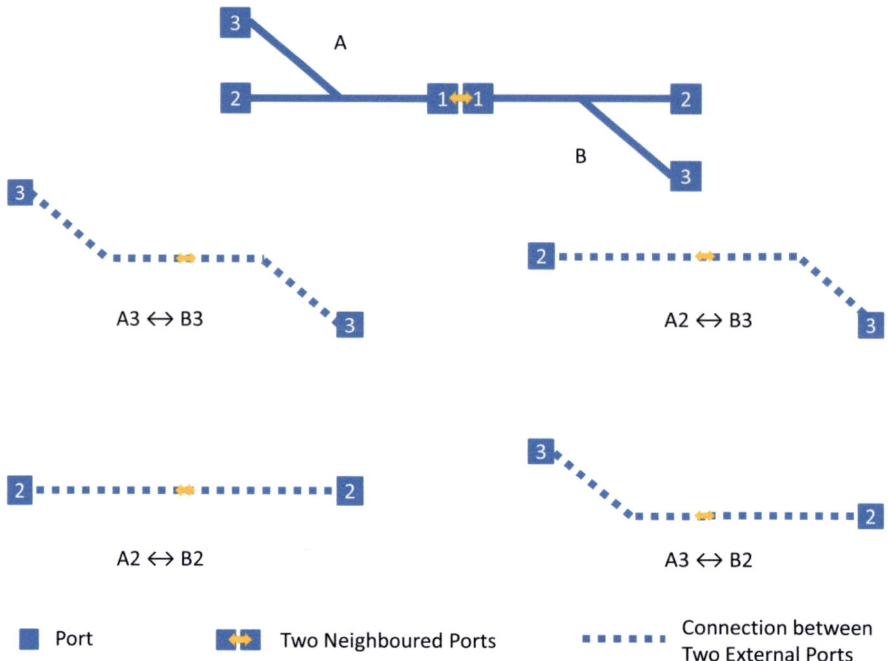

Figure A 1: An example with all overlapped connections

Given a junction-type basic structure R with turnouts A and B, the overlapping situation is further examined for the additional turnout C in Figure A 2. The examined fictive junction-type basic structure R' includes turnouts A, B, and C. The external ports of R' are A2, A3, B3, C1, and C3. For the connection A3↔B3 and the connection C3↔C1, there is no overlapping situation. It means that two trains can run through the A3↔B3 and C3↔C1 simultaneously. Therefore, the turnout C cannot be assigned to the junction-type basic structure R.

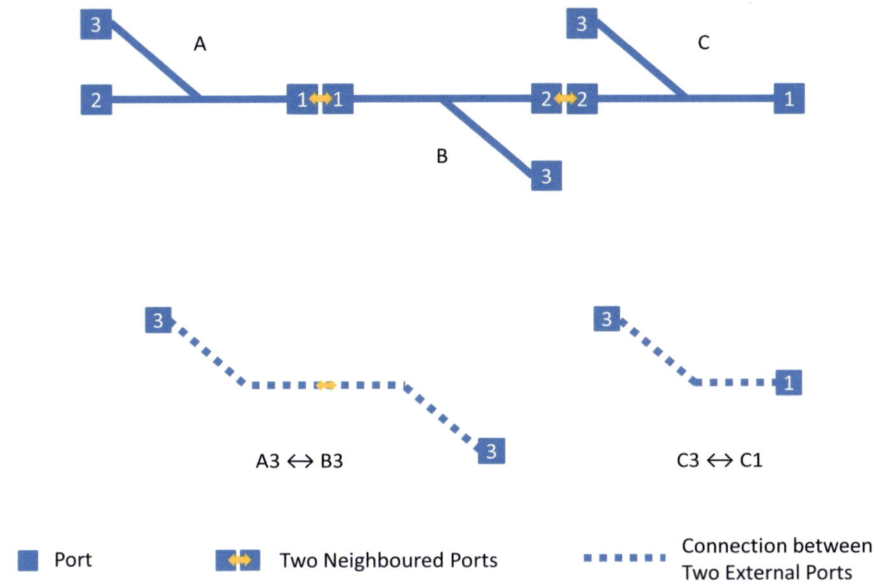

Figure A 2: An example of two connections without overlapping

The activity diagram for establishing junction-type basic structures is shown in Figure A 3. Once each junction-type infrastructure element has been assigned to a junction-type basic structure, station tracks can be established for the tracks, which have not yet been assigned to junction-type basic structures. The tracks inside a station tracks connect the identical junction-type basic structures. A station tracks are connected to either one (for the dead-end tracks) or two junction-type basic structures. Through comparing the connected junction-type basic structures, each track can be assigned to a station tracks respectively.

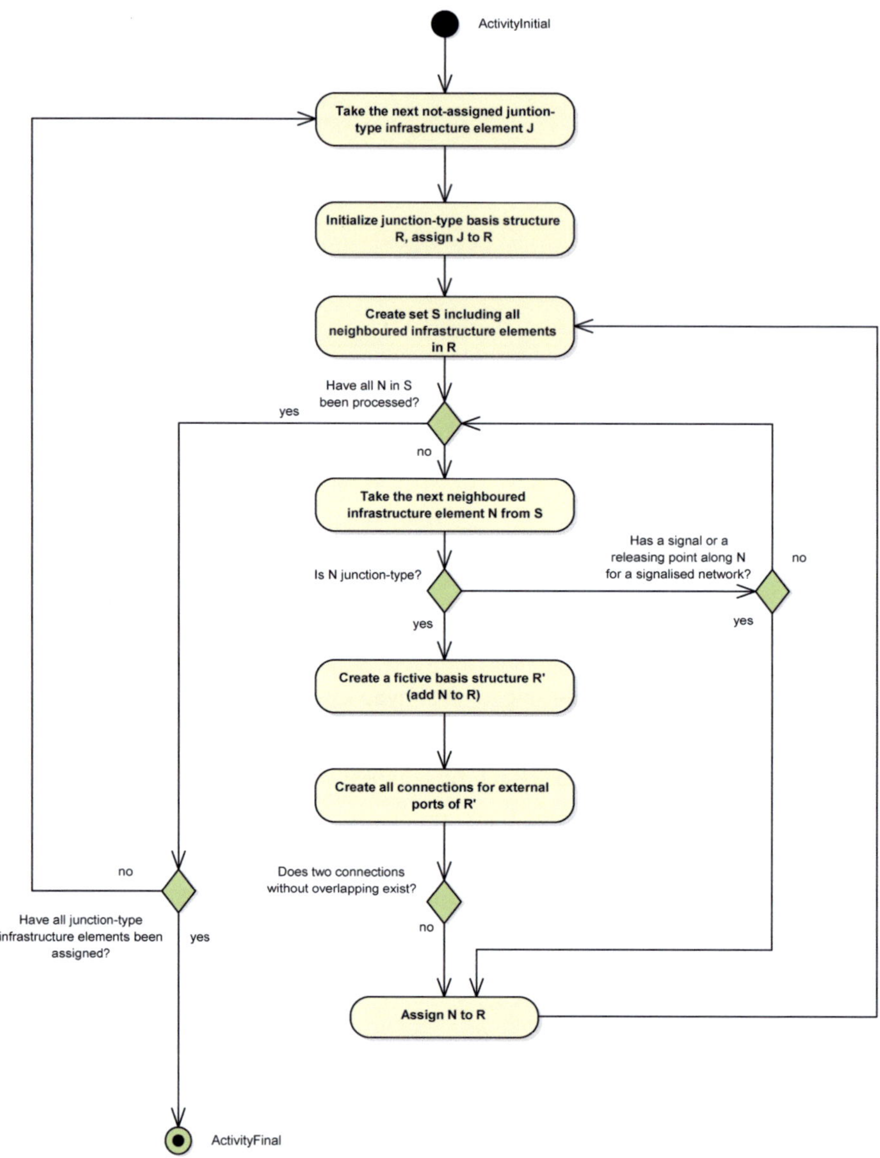

Figure A 3: Workflow of establishing junction-type basic structures

Appendix B: Calculation of Running Dynamics with Minimum Energy Consumption

The algorithm for the calculation of running dynamics with minimum energy consumption is implemented in the software PULSim. In Section 4.1.2, the sections of a train run are defined. A moving section represents an independent part for the calculation of running dynamics inside of an authorised running section. Several characteristic sections are included in a moving section. In each characteristic section, there are five behaviour sections: acceleration, cruising, coasting, and braking.

The theory of optimal control for railway operations and control was initially published in (Strobel et al. 1974). The derivation of the optimal driving regimes are further developed and presented in (Howlett, Pudney 1995), (Liu, Golovitcher 2003) and (Albrecht 2014). Based on the theory of optimal control, five optimal driving regimes are applied depending on the running dynamics of different behaviour sections:

- Acceleration: The train should accelerate with maximal acceleration rate.
- Cruising with the resistances against the running direction: The train should provide tractive force to overcome the resistance.
- Coasting: The tractive force and the braking force are zero.
- Cruising (the direction of the resistance is same as the running direction): The train should provide braking force to balance the resistance.
- Braking: The braking force should be provided with maximal braking rate.

For a given timetable, recovery time has been already scheduled. The recovery time can be calculated through comparing the difference between the regular running time and the minimum running time (see Section 7.2.2). The key to implementing a driving style with minimum energy consumption is to distribute the recovery time to the characteristic sections, so that the behaviour sections for each characteristic section can be determined. In PULSim, the approach developed by (Horn, Zinke) is applied.

The minimum running time for a moving section should be calculated at first in order to derive the available recovery time. The time reserve will be distributed to each characteristic section iteratively. For a characteristic section, the initial value of the maximum velocity is set as the maximum velocity for the minimum running time. Several rounds of running time calculation are carried out with the stepwise reduced maximum velocity. The saved energy ΔE and the extension of running time Δt for each step of the maximum velocity will be saved accordingly. The characteristic sections will be ordered by the value of $\frac{\Delta E}{\Delta t}$ in descending order. The characteristic section with the maximal value of $\frac{\Delta E}{\Delta t}$ will be chosen. The running time of the characteristic section will be extended by Δt. The process will be repeated until the available recovery time has been distributed. Since the maximum speed of a characteristic section is reduced in stepwise (e.g., by 5 km/hour), the sum of the extended running time may exceed the available recovery time at a step of the maximum velocity. The process of examining the characteristic section will be terminated before the step with a certain tolerance of errors.

This algorithm is based on velocity-step approach. The errors between the achieved running time and the scheduled running time can't be eliminated completely. In future development, a combination of time-step approach can be applied to distribute the recovery time exactly.

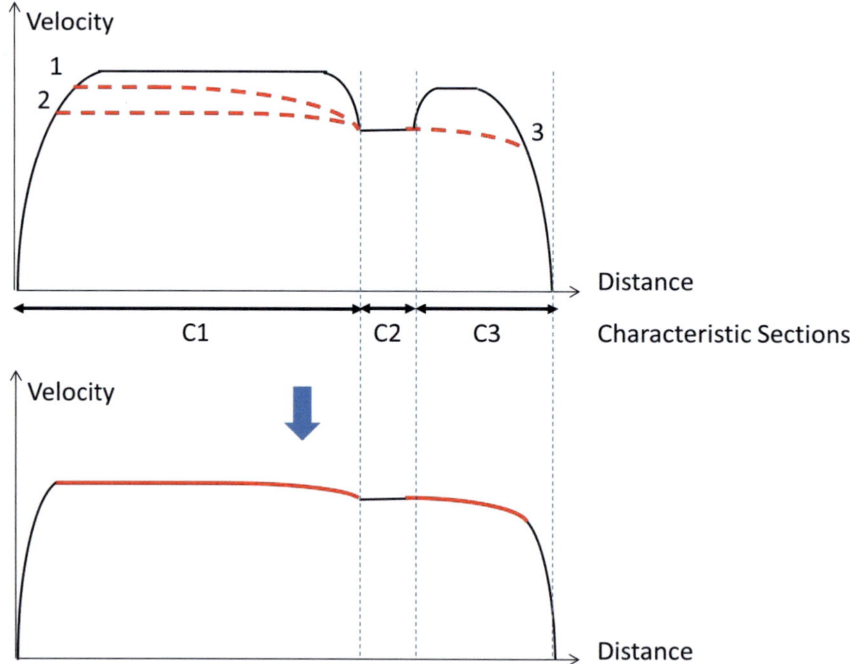

Figure B 1: Distributing recovery time in a moving section

An example of distributing recovery time in a moving section is illustrated in Figure B 1. The moving section includes three characteristic sections C1, C2, and C3. For the first and second rounds of distribution, the recovery time are applied on the characteristic section C1, which reaches the highest value of $\frac{\Delta E}{\Delta t}$ among all the characteristic sections. For the third round of distribution, the available recovery time is not enough for C1. Therefore, the running time is extended at the characteristic section C3.

In Table B 1, the calculated running dynamics in PULSim shown in Figure 4-9 are listed in form of discrete points (the definition of discrete point is given in Section 4.1.1):

No.	Distance (m)	Velocity (km/h)	Duration (s)
1	0,000	0,000	0,000
2	0,000	0,000	192,000
3	0,000	0,000	3,799
4	0,000	0,000	0,000
5	0,965	5,000	1,389
6	2,894	10,000	1,389
7	4,823	15,000	1,389
8	6,752	20,000	1,389
9	8,681	25,000	1,389
10	10,610	30,000	1,389
11	12,539	35,000	1,389
12	14,468	40,000	1,389
13	16,397	45,000	1,389
14	18,326	50,000	1,389
15	20,255	55,000	1,389
16	23,466	60,000	1,469
17	27,965	65,000	1,611
18	32,903	70,000	1,755
19	38,162	75,000	1,895
20	43,768	80,000	2,033
21	0,000	80,000	0,000
22	96,030	80,000	4,321
23	280,799	80,000	12,636
24	107,201	79,182	4,980
25	0,000	79,182	0,000
26	323,693	79,182	14,717
27	548,685	75,000	25,622
28	658,498	70,000	32,698
29	38,737	65,000	2,066
30	36,006	60,000	2,074
31	33,247	55,000	2,082

32	30,462	50,000	2,089
33	27,651	45,000	2,096
34	24,817	40,000	2,102
35	21,961	35,000	2,108
36	19,085	30,000	2,114
37	16,190	25,000	2,119
38	13,277	20,000	2,124
39	10,349	15,000	2,129
40	7,406	10,000	2,133
41	4,451	5,000	2,137
42	1,486	0,000	2,140
43	0,000	0,000	30,000
44	0,000	0,000	0,000
45	0,965	5,000	1,389
46	2,894	10,000	1,389
47	4,823	15,000	1,389
48	6,752	20,000	1,389
49	8,681	25,000	1,389
50	10,610	30,000	1,389
51	12,539	35,000	1,389
52	14,468	40,000	1,389
53	16,397	45,000	1,389
54	18,326	50,000	1,389
55	20,255	55,000	1,389
56	23,466	60,000	1,469
57	27,965	65,000	1,611
58	32,903	70,000	1,755
59	533,404	70,000	27,432
60	54,557	69,586	2,910
61	0,000	69,586	0,000
62	3,022	70,000	0,156
63	529,422	70,000	27,227
64	659,381	65,000	35,167
65	657,793	60,000	37,889
66	33,247	55,000	2,082
67	30,462	50,000	2,089
68	27,651	45,000	2,096
69	24,817	40,000	2,102
70	21,961	35,000	2,108
71	19,085	30,000	2,114
72	16,190	25,000	2,119
73	13,277	20,000	2,124
74	10,349	15,000	2,129
75	7,406	10,000	2,133
76	4,451	5,000	2,137

77	1,486	0,000	2,140
78	0,000	0,000	30,000
79	0,000	0,000	0,000
80	0,965	5,000	1,389
81	2,894	10,000	1,389
82	4,823	15,000	1,389
83	6,752	20,000	1,389
84	8,681	25,000	1,389
85	10,610	30,000	1,389
86	12,539	35,000	1,389
87	14,468	40,000	1,389
88	16,397	45,000	1,389
89	18,326	50,000	1,389
90	20,255	55,000	1,389
91	23,466	60,000	1,469
92	27,965	65,000	1,611
93	32,903	70,000	1,755
94	38,162	75,000	1,895
95	43,768	80,000	2,033
96	49,856	85,000	2,176
97	56,601	90,000	2,329
98	423,166	90,000	16,927
99	645,529	85,000	26,559
100	651,203	80,000	28,416
101	44,111	75,000	2,049
102	41,439	70,000	2,058
103	38,737	65,000	2,066
104	36,006	60,000	2,074
105	33,247	55,000	2,082
106	30,462	50,000	2,089
107	27,651	45,000	2,096
108	24,817	40,000	2,102
109	21,961	35,000	2,108
110	19,085	30,000	2,114
111	16,190	25,000	2,119
112	13,277	20,000	2,124
113	10,349	15,000	2,129
114	7,406	10,000	2,133
115	4,451	5,000	2,137
116	1,486	0,000	2,140
117	0,000	0,000	30,000
118	0,000	0,000	0,000
119	0,965	5,000	1,389
120	2,894	10,000	1,389
121	4,823	15,000	1,389

122	6,752	20,000	1,389
123	8,681	25,000	1,389
124	10,610	30,000	1,389
125	12,539	35,000	1,389
126	14,468	40,000	1,389
127	16,397	45,000	1,389
128	18,326	50,000	1,389
129	20,255	55,000	1,389
130	23,466	60,000	1,469
131	27,965	65,000	1,611
132	32,903	70,000	1,755
133	38,162	75,000	1,895
134	43,768	80,000	2,033
135	49,856	85,000	2,176
136	56,601	90,000	2,329
137	63,989	95,000	2,490
138	72,107	100,000	2,662
139	80,925	105,000	2,842
140	90,401	110,000	3,027
141	100,547	115,000	3,217
142	0,000	115,000	0,000
143	54,650	117,484	1,693
144	1494,954	117,484	45,809
145	64,863	117,484	1,988
146	31,633	115,000	0,980
147	61,878	110,000	1,980
148	59,448	105,000	1,991
149	56,979	100,000	2,001
150	54,475	95,000	2,011
151	51,935	90,000	2,021
152	49,360	85,000	2,031
153	46,752	80,000	2,040
154	44,111	75,000	2,049
155	41,439	70,000	2,058
156	38,737	65,000	2,066
157	36,006	60,000	2,074
158	33,247	55,000	2,082
159	30,462	50,000	2,089
160	27,651	45,000	2,096
161	24,817	40,000	2,102
162	21,961	35,000	2,108
163	19,085	30,000	2,114
164	16,190	25,000	2,119
165	13,277	20,000	2,124
166	10,349	15,000	2,129

167	7,406	10,000	2,133
168	4,451	5,000	2,137
169	1,486	0,000	2,140
170	0,000	0,000	60,000
171	0,965	5,000	1,389
172	2,894	10,000	1,389
173	4,823	15,000	1,389
174	3,113	17,484	0,690
175	1,089	17,484	0,224
176	4,774	15,000	1,058
177	7,406	10,000	2,133
178	4,451	5,000	2,137
179	1,486	0,000	2,140

Table B 1: The discrete points of running dynamics for train run shown in Figure 4-9